KU-655-119

SCHRÖDINGER'S CAT

AND 49 OTHER EXPERIMENTS THAT REVOLUTIONISED PHYSICS

ADAM HART-DAVIS

(m)

Contents

Introduction

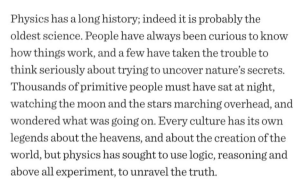

Physics has a long history; indeed it is probably the oldest science. People have always been curious to know how things work, and a few have taken the trouble to think seriously about trying to uncover nature's secrets. Thousands of primitive people must have sat at night, watching the moon and the stars marching overhead, and wondered what was going on. Every culture has its own legends about the heavens, and about the creation of the world, but physics has sought to use logic, reasoning and above all experiment, to unravel the truth.

Astronomy has always been at the forefront of science; you can observe the heavens with your naked eyes, and compile lists and atlases of stars, noting the curious movements of the planets and the occasional appearance of shooting stars, comets and supernovae. The invention of the telescope around 1600 allowed astronomy to move up a step, but astronomers did not conduct experiments, which is why there are few of them in this book.

The difference between Empedocles' experiment with a bucket, and Archimedes' revelation in the bath, was a gap of 200 years, and a big leap forward in calculation and understanding. After the Greek civilization faded there was a period of little advance, until the dawning of the Golden Age of Islam, when many Arab scientists, engineers and alchemists took science further. But then came another fallow period, until Copernicus produced (in 1543) a book about the sun-centred universe, and Galileo was persuaded to agree with him 67 years later, after seeing the moons of Jupiter.

Galileo undertook a series of groundbreaking experiments, and then along came Robert Boyle and Isaac Newton to put physics and chemistry on firm foundations.

Armed with new practical and theoretical skills, scientists measured the speed of sound, the speed of light, the mass of Earth and the aerodynamics of wings. The great majority of this work was done in Europe, and especially in Germany, but then Americans began to make their mark, and have done so ever since.

Towards the end of the 19th century there was a flurry of amazing discoveries – x-rays, radioactivity and electrons all appeared within five years, and these led to further ideas, further theories and further experiments; the early years of the 20th century saw an extraordinary advance in the understanding of the nature of matter.

The two world wars hijacked researchers to work on military projects, out of which came radar and microwaves, tokamaks, and nuclear power. After the wars, however, fundamental science blossomed once more, and in particular astronomers, astrophysicists and space scientists began to probe more deeply into the nature of the universe. Telescopes were put into space, where there is no atmosphere to spoil the view of the stars, and computing power grew according to Moore's Law, which states that the number of transistors in a dense integrated circuit (and therefore the power of computers) doubles every two years.

The 21st century is the era of Big Science, involving the largest and most expensive experiments ever tried, some including thousands of physicists, and harnessed to many supercomputers to analyse the massive streams of data they generate.

Yet, however many experiments are done, each one always raises new questions; new questions waiting for answers.

Adam Hart-Davis

CHAPTER 1: Early experiments:
430 BCE – AD 1307

The ancient Chinese were great inventors, coming up with such wonders as the magnetic compass, gunpowder, paper, printing and Zhang Heng's superb seismometer, which detected a distant earthquake. They also had great astronomers, who spotted a supernova as far back as 1054.

The ancient Greeks were more interested in science generally, and Aristotle in particular wrote at length about physics, biology, zoology and other sciences. Aristotle did not do practical experiments himself, but

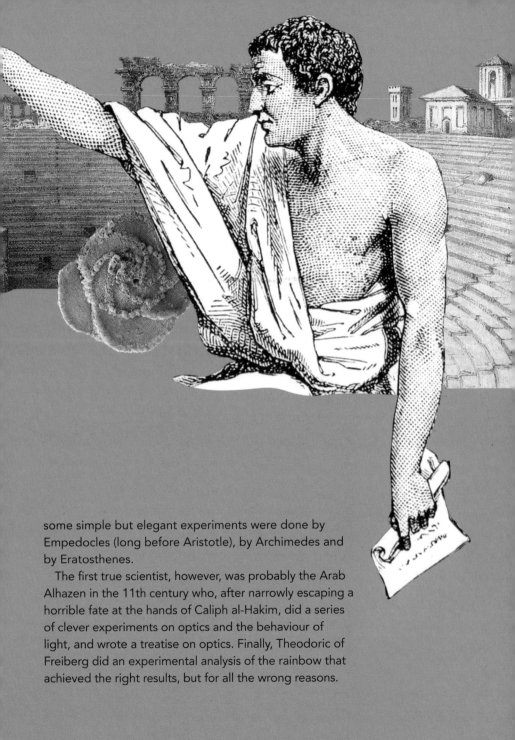

some simple but elegant experiments were done by Empedocles (long before Aristotle), by Archimedes and by Eratosthenes.

The first true scientist, however, was probably the Arab Alhazen in the 11th century who, after narrowly escaping a horrible fate at the hands of Caliph al-Hakim, did a series of clever experiments on optics and the behaviour of light, and wrote a treatise on optics. Finally, Theodoric of Freiberg did an experimental analysis of the rainbow that achieved the right results, but for all the wrong reasons.

ca 430 BCE

THE STUDY

RESEARCHER:

Empedocles

SUBJECT AREA:

Pneumatics

CONCLUSION:

Air is a material
substance.

IS AIR 'SOMETHING'?

EMPEDOCLES' SEARCH FOR THE
ROOTS OF EVERYTHING

The town of Agrigento, in the middle of the southwest coast of the island of Sicily, has some of the finest Greek temple remains anywhere; they stand end to end on a high ridge, proud in the sunshine. There is also a glorious amphitheatre. In the 5th century BCE there lived in the town a Greek philosopher called Empedocles, who performed one of the earliest known scientific experiments, in order to back up his theory about the elements.

Four elements

For hundreds of years others had pondered and argued about what things were made of. Thales had suggested water, because it can turn into ice and into steam; so perhaps it can turn into anything. Others had suggested various combinations of substances. Empedocles declared that everything was a mixture of the four elements (or 'roots', as he called them) Earth, Air, Fire and Water, simply combined in various proportions.

Each element, he said, wants to get back to where it belonged; so earth always falls downwards; water trickles to the sea; air bubbles up through water; and fire tries to get up to the sun.

These elements were unchanging. They were bound together by love, but always being torn apart by strife; so they were in a constant state of flux.

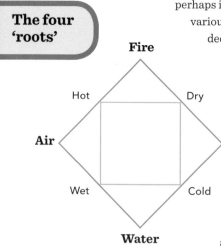

The four
'roots'

Fire

Hot — Dry

Air — Earth

Wet — Cold

Water

But there was a bit of a problem, because some cynics said that air could not be an element. Air was nothing; it could not be part of anything; it could not be a root. Empedocles pointed out that air bubbles up through water.

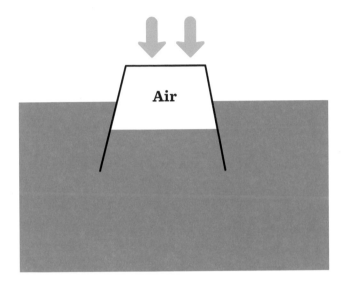

You can see the bubbles; they must be something. This did not satisfy his critics; so he devised a cunning experiment.

Submerging the water clock

To tell the time he used a clepsydra – a water clock – which was a ceramic jar with a small hole in the bottom to let the water trickle out. He turned his clepsydra upside down, put his finger over the hole, and thrust it down into the sea. When he took it out again, he showed that the inside was still dry at the bottom, demonstrating that something had kept the water out. This something had to be air, and therefore air was indeed something, and not nothing.

The idea of the four elements Earth, Air, Fire and Water was not seriously challenged until Robert Boyle redefined the notion of the element more than 2,000 years later.

A fiery ending

Empedocles believed he was immortal, and to prove this to his followers he led them up to the top of Mt Etna, the active volcano that towers over the eastern end of the island. There, he is said to have jumped into the smoking crater.

According to one legend, one of his sandals was ejected, but he was never seen again. This precipitate action sounds like a mistake, but as a result we remember him to this day; so perhaps it was a good way to become immortal.

WHY DOES THE BATH OVERFLOW?

ARCHIMEDES' EUREKA MOMENT

ca 240 BCE

THE STUDY

RESEARCHER:
Archimedes

SUBJECT AREA:
Hydrostatics

CONCLUSION:
The discovery of
the displacement
of liquids.

Archimedes lived in Syracuse on the island of Sicily from about 287 BCE. until 212, when he was killed by a Roman soldier during the invasion of the city. He was the finest mathematician of the ancient world, and the thing he was most proud of was proving, without the aid of the equations we have today, that if a sphere just fits inside a closed cylinder, like an orange in a small tin, then the volume of the sphere is two-thirds of the volume of the cylinder, and the surface area is also two-thirds that of the cylinder. He asked that his tombstone should be carved with a diagram of a sphere inside a cylinder, and it was duly discovered by the Roman orator Cicero 137 years later.

War machines

Archimedes was also a skilful engineer, and when the Roman fleet came to invade in 212 he organized all sorts of defensive machines, which apparently included catapults, cranes to lift one end of a boat out of the water and so

sink it, and a death ray: a large squad of soldiers were to angle all their polished shields to reflect the sun's rays on to an approaching ship and so set it on fire.

He also worked out the laws of levers and pulleys, managed to move a fully laden ship with the aid of a bunch of pulleys, and famously said 'Give me a lever that is long enough, and a place to stand, and I will move the world.'

The suspect crown

Archimedes' greatest triumph, however, was solving the riddle of the suspect crown. King Hieron II was a tyrant who had ordered a new crown from the royal crown maker, and had given him a lump of pure gold (just over 1 kilogram or 2 pounds) to make it with. When the fancy crown appeared, Hieron suspected that the man had stolen some of the gold and substituted an equal weight of silver. So the weight was still 1 kilogram, but was it all gold? Hieron sent for Archimedes and asked him to find out.The challenge was difficult. The crown was highly elaborate, and Archimedes was not allowed to damage it in any way. While he was pondering the problem, Archimedes went to have one of his rare baths in the public bathhouse in town.

The all-important bath

As Archimedes stepped in, he noticed two things: first, the water level rose a little as he immersed his body, and some water slopped over the side of the bath. Second he felt light, almost weightless. At this point he had a flash of inspiration, and according to legend he leapt out of the bath, shouting 'Eureka' (meaning 'I have found it' or 'I have the answer') and ran all the way home stark naked.

The two important things he realized were:
1. When a body sinks into water it has to push some water out of the way – it displaces the water.
2. Any body in water feels light, because it feels an upthrust equal to the weight of water displaced. This is now known as the Archimedes' Principle.

In theory he could have filled a bucket to the brim, lowered the crown in and measured the amount of water that was displaced and spilled over the edge. That would give him the volume of the crown, from which he could have worked out its density, since density is equal to mass divided by volume.

He knew the volume of 1 kilogram of pure gold would be 52 cubic centimetres (20 cubic inches), and he would expect the volume to be greater if there was some silver mixed in, because the density of silver is less than that of gold – so the volume of silver would be greater than the volume of the same mass of gold.

Using Archimedes' Principle

Measuring volumes accurately is difficult, however; so what he probably did was use his upthrust idea. He borrowed just over 1 kilogram of pure gold from the king, hung the crown and the new lump on a beam so that they balanced, and then lowered the whole thing into a tub of water. If the crown was impure, then its volume would be greater than 52 cubic centimetres, and so it would experience a greater upthrust, because upthrust depends on volume. So he would expect the crown to float upwards in the water.

Sure enough, the crown did float upwards. The maker had substituted some other metal, and was duly punished.

Archimedes wrote a number of books, or treatises, of which a dozen have survived. One was *On the Sphere and the Cylinder;* another *On Floating Bodies* and another, *The Sand Reckoner,* was a huge calculation of how many grains of sand it would take to fill the universe, for which he had to invent an entirely new set of numbers.

HOW CAN WE MEASURE THE EARTH?

SUN, SHADOWS AND EARLY GREEK GEOMETRY

ca 230 BCE

THE STUDY

RESEARCHER:
Eratosthenes

SUBJECT AREA:
Geometry

CONCLUSION:
The circumference
of Earth is approx.
40,000 kilometres
(25,000 miles).

The Greek city of Alexandria, at the mouth of the Nile in Egypt, was founded by Alexander the Great in 322 BCE. He ordered the construction of a harbour by making a breakwater to the little island of Pharos, just offshore. There, he said, should stand a great lighthouse, which came to be called the Pharos, and was one of the seven wonders of the ancient world. In the third century BCE, Alexandria became the centre of learning in the Greek world, with a magnificent library containing hundreds of thousands of scrolls made of parchment or vellum. Around 240 BCE, a new librarian was appointed – Eratosthenes, a mathematician who devised a method of finding prime numbers. It's known as Eratosthenes' Sieve.

Prime numbers

Suppose you want to find all the prime numbers between 2 and 50 (1 is not generally counted as a prime). Write them all down in a grid. Cross out all the even numbers higher than two, because they are divisible by two. Cross out all the numbers higher than three which are divisible by three. Then repeat for five and seven, and you are left with all the primes up to 50: 2, 3, 5, 7, 11, 13, 17, 19, 23, 29, 31, 37, 41, 43 and 47.

Sizing the world

Eratosthenes was also a geographer, indeed probably the finest in the ancient world. The ancient Greeks knew that

the Earth was round; they had two solid pieces of evidence. First, when a ship sailed away from the shore, it gradually disappeared, from the bottom up. Quite soon the hull was lost to view, and then the masts. Clearly it was not just getting too small to see, but was going over the horizon, which meant that Earth was a sphere. Second, they realized that an eclipse of the moon was caused by the shadow of Earth, and this shadow was curved.

Knowing that Earth was a sphere, Eratosthenes wanted to find out how big it was. Eight hundred kilometres (500 miles) south of Alexandria, at Syene (now Aswan) there is a well on Elephant Island in the Nile. Eratosthenes knew that at noon on midsummer's day anyone who looked down the well could see a reflection of the sun. This had to mean that the sun was exactly overhead at that time. The well is still there, but unfortunately it is now dry and full of rubble, and there are no reflections.

Angling the sun

Back in Alexandria, Eratosthenes planted a stick vertically in the ground, and at noon on midsummer's day measured the angle of the sun, or rather the angle between the stick and the edge of its shadow; it was 7.2 degrees. This is the angle A in the diagram on page 19. This was the same as the angle A*, because they are on either side of a diagonal between parallel lines. A* is the angle at the centre of Earth between Alexandria and Syene; so Eratosthenes was able to make this simple calculation:

- Angle between Alexandria and Syene = 7.2 degrees
- Distance from Alexandria to Syene = 500 miles
- Angle around Earth from Alexandria to Alexandria
  = 360° = 50 x 7.2 degrees

Therefore distance around Earth = 50 x 500 = 25,000 miles.

The distance from Alexandria to Syene had been measured by official *bematistoi* (surveyors who had been trained to pace evenly and count their paces), and Eratosthenes gave the result in 'stades', rather than miles. We don't know exactly how long the stade was, but as far as we know his estimate of the circumference of Earth was close to today's accurate value of 40,075 kilometres (24,900 miles).

Archimedes travelled from Sicily to meet his friend Eratosthene in Egypt, and may have invented the Archimedes screw while he was there; Archimedes screws are still used to pump water from the Nile for irrigation.

Archimedes used to send Eratosthene postcards (or the Greek equivalent) with enormously complicated mathematical puzzles. One was about a vast herd of cows and bulls each of four different colours; the problem was to work out from a set of equations how many there were in each group. One of the answers was such a huge number it needed more than 200,000 digits.

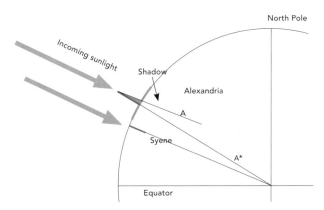

AD 1021

THE STUDY

RESEARCHER:
Alhazen

SUBJECT AREA:
Optics

CONCLUSION:
Light travels in straight lines.

HOW DOES LIGHT TRAVEL?

THE INVENTION OF THE CAMERA OBSCURA

One of the first scientists to carry out systematic experiments was the Arab Abū Alī al-Hasan ibn al-Hasan ibn al-Haytham, or ibn al-Haytham for short, but Latinized to Alhazen (or Alhacen).

Alhazen was born in Basra, in modern-day Iraq, in 965, and educated in Baghdad. In his forties, he heard about the problems in Egypt caused by the annual flooding of the Nile, and he rashly wrote to Caliph al-Hakim, offering to sort it out. The Caliph welcomed him to Cairo with great ceremony, and sent him up the Nile to do the job.

Alhazen's plan to build a dam at what is now Aswan was entirely sensible, but he had not appreciated the scale of the problem. When he had travelled way down south to Aswan, he found the river was a mile wide, albeit split into several streams. He could not possibly build a dam with the technology that was available, but he was terrified of admitting his failure, because he was sure that the tyrannical and unforgiving Caliph would cut his head off. So instead he decided to 'go mad', and he remained 'mad' and under house arrest for ten years, until the Caliph died in 1021.

Studying the working of the eye

Alhazen used those ten years to study optics, carrying out
a series of experiments. First he thought about how the eye
works. Euclid, Ptolemy and others had said that in order
to see something – for instance a tree – we open our eyes
and send out a beam of light to illuminate the tree; the light
then bounces back into our eyes to make an image. Aristotle
thought that actual shapes came into our eyes.

Alhazen said that this was all nonsense. The light is
out there anyway. During the day everything is lit by the
sun, and the sunlight bounces off trees, houses and people
and into our eyes. As he said, 'from each point of every
coloured body, illuminated by any light, issue light and
colour along every straight line that can be drawn from
that point.' All we have to do is open our eyes and the
light floods in. He dissected bulls' eyes to find out what
was inside, and drew excellent diagrams to explain the
structure of the human eye, and how it worked.

He said that the moon looks bigger when it is near
the horizon because there are trees and other objects to
compare it with, and it seems to be far away. When it is
high in the sky and alone, it looks nearer, but as a result it
appears to be smaller.

The camera obscura

Alhazen suspected that light always travels in straight
lines, because objects in sunlight cast shadows with
fairly sharp edges. In order to demonstrate this
convincingly he built a camera obscura, which means
'dark room'. This was a small dark room, with a little
hole in the shutters on one side, opposite a white wall or
screen. The bright Egyptian sun illuminated the world
outside, and light came streaming through the small hole
to cast an image on the opposite wall. This image was
upside down and reversed right to left, but nevertheless
clearly showed the world outside, moving and in colour.

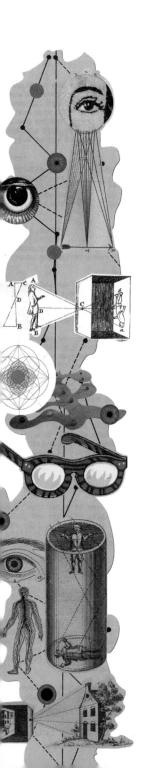

People who saw this were amazed. They had never seen images like it before.

Alhazen explained that in order to form the image, light from the outside world must be coming through the hole in straight lines; otherwise the image would just be a great jumbled splurge of colour.

He set up his camera obscura in the evening, when it was dark, and the only light from outside came from three hanging lamps. Inside, three spots of light appeared on the wall, one from each lamp. He showed that the light came through the hole and in a straight line to each spot. He could intercept the light going to any one spot by just putting his hand in the way and catching it. This was the clearest demonstration yet that light travels in straight lines.

The book of optics

Alhazen also experimented with lenses, mirrors, reflection and refraction, and described his theories and experiments in his *Book of Optics* or *Kitab al-Manazir*. This was one of the first books of experimental science and was admired, centuries later, by Leonardo da Vinci, Galileo, Descartes and Isaac Newton. In all he wrote more than 200 books, of which some 50 have survived.

Most of all, however, Alhazen could be said to be the first scientist. He has been called the founder of the scientific method, being sceptical of other writers, and relying on the systematic observation of physical phenomena and their relationship to theory:

> *The duty of the man who investigates the writings of scientists, if learning the truth is his goal, is to make himself an enemy of all that he reads, and. . . attack it from every side. He should also suspect himself as he performs his critical examination of it, so that he may avoid falling into either prejudice or leniency.*

WHY IS THE RAINBOW RAINBOW-COLOURED?

UNDERSTANDING THE PATHWAYS OF LIGHT

ca 1307
THE STUDY
RESEARCHER:
Theodoric of Freiberg
SUBJECT AREA:
Optics
CONCLUSION:
Rays of light travel
on paths.

Theodoric was born in Germany before 1250 and became a Dominican friar, rising to the high office of Provincial of Germany from 1293 to 1296. At the General Chapter at Toulouse in 1304, Aymeric, the Master General, suggested that Theodoric should make a scientific study of the rainbow.

Theodoric was an independent man, not inclined to follow conservatively the established order and doctrines. For example he preached in German, rather than Latin. His independent outlook meant that he took investigations both seriously and scientifically, and used experiments rather than hearsay at every possible stage of the argument.

A mistaken theory of colour

Theodoric's explanation of the colours was original, verified by experiment and entirely wrong. He did not think about a continuous spectrum (red-orange-yellow-green-blue-indigo-violet) as we do now, but believed there were four dominant colours – red, yellow, green and blue – of which red and yellow were 'clear' or translucent colours, while blue and green were 'obscure' or opaque.

Furthermore, he believed that if light was moving near the edge of a piece of glass or near the surface of water, the clear colour would be red, but in the centre, away from the boundary, it would be yellow. Also if the material were transparent the obscure colour would be green, but if it were opaque the obscure colour would be blue.

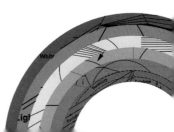

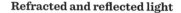

Refracted and reflected light

Theodoric tested these ideas by passing sunlight through a glass prism. He expected clear colours to be refracted near the surface, and obscure colours deeper inside the glass; also that the clear colour red would be nearest to the surface, and the obscure colour blue would be furthest away, since the prism would be most opaque in the centre. Therefore he expected the colours in the order red, yellow, green, blue.

When he peered through his hexagonal prism at the sun, or let the light pass through and fall on a screen, he saw that indeed the colours were in the order he had predicted. His diagram shows that he understood that the light had been refracted twice, both on entering the prism and on leaving it, and that the colours had come from within the prism. In addition it shows that the light may also be reflected inside the prism.

The pathways of light

Theodoric then took a large round glass flask and filled it with water, so as to make a model of a raindrop. Then he peered at the sun through the flask, and as he moved his head up and down he saw the same order of colours, but now they were reversed; red was at the top, and blue at the bottom, exactly as they are in a rainbow. He realized that this reversal was caused by the fact that the rays of light were reflected inside the flask, as well as being refracted twice. This is clear from his diagrams.

He therefore showed that light rays of particular colours travel on specific pathways through the flask, and that these pathways produced the colours; they were not merely in the eye of the beholder.

He then proposed that the paths of sunlight through raindrops were just the same as through his flask of water, on the assumption that they fall so fast, and are replaced so rapidly by others, that they are equivalent to a stationary curtain of water drops.

Unfortunately this diagram suggests that the sun is only about as far from the observer as the raindrops, which means that its rays are not parallel. Nevertheless the principle is good, and it does help to explain why the bow is circular.

In fact the sun is enormously far away. Imagine a line from the sun through your head to the ground – to the shadow of your head. The rainbow is always formed at 42 degrees to this line. So the maximum height is 42 degrees – if the sun is on the horizon – and the rainbow always forms part of a circle. If you see one from an aircraft or from a mountain you may see a complete circular rainbow.

You cannot get to the end of the rainbow because there is no physical object – merely this arc of a circle in the sky – and it moves when you move.

The reversal
When Theodoric held his glass flask at the correct angle he saw the secondary rainbow, in which the colours are reversed, with blue at the top. This time he understood that the rays had been reflected twice within the water drops.

Although his theory of refraction and the colours, and his measurement of the angle of the rainbow, were hopelessly wrong, Theodoric thus set a marvellous example of the use of models and the scientific method – proposing a theory and then testing it by experiment.

CHAPTER 2: The Enlightenment: 1308-1760

Throughout the dark ages it seems that even philosophers were inclined to accept religious doctrine: 'Why did this phenomenon happen?' Answer: 'It was the will of God.' Then a few people began to look for more logical explanations, and to test their ideas by experiment. British philosopher Francis Bacon wrote books in the 1620s that encouraged empirical evidence and experimental science.

Robert Norman and Galileo had already taken up the cudgel of experiment, and more followed. Isaac Newton displayed his powerful intellect in his first scientific paper, and various researchers investigated the speed of light, the speed of sound and the heat hidden in melting ice. Above them all towers his mighty tome, *Principia Mathematica* (1687).

1581

THE STUDY

RESEARCHER:

Robert Norman

SUBJECT AREA:

Earth science

CONCLUSION:

A free-floating
compass needle dips
sharply downward
towards the pole.

WHERE IS
MAGNETIC NORTH?

CHASING THE COMPASS NEEDLE

In the middle years of the 16th century, Robert Norman spent many years at sea and then settled in England, near London, and became an instrument maker, especially of compasses, for the compass was the sailor's most important instrument for navigation. He made his compass needles from iron, and then magnetized them by stroking them with a chunk of loadstone (or lodestone), a naturally occurring magnetic rock called magnetite.

He knew all about magnetic variation – a compass needle does not always point due north – but then he discovered magnetic dip or 'declination' as he called it, and gives a vivid description of how he was persuaded to investigate it.

He noticed that even his best compass needles, when balanced on a fine point, would not only turn towards north, but would tilt, so that he had to put a counterweight on the south end to keep them level. One day, after making a very fine needle and pivot, he found the declination was strong; so he began to shorten the north end to reduce it. He wrote:

*In the end I cut it too short, and so
spoiled the needle wherein I had taken such*

*pains. Hereby being strocken into some choler [anger],
I applied myself to seek further into this effect.*

Norman decided to make what we now call a dip
circle, to investigate the effect, but first he wanted to
find out the cause of the dip; was it just magnetic, or
had the north end of the needle absorbed from the
loadstone 'some ponderous and weighty matter'?

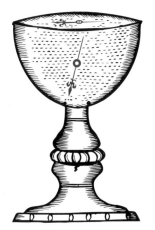

The first compass

He put some small pieces of iron in a balance pan,
and balanced them with some pieces of lead. Then he
magnetized the iron with his loadstone and put the
pieces back in the pan. He reported:

*You shall find them to weigh no more, than before
they were touched. Furthermore if the north end
of the needle had taken up something weighty from
the loadstone, so too the south end should have taken
up something weighty from the other end of the
loadstone, and there would be no dip effect.*

The wine glass experiment

*Now you shall take a piece of iron or steel wire of five
centimetres long or more, and thrust it into a piece of
close cork, as big as you think may sufficiently bear the
wire on the water, so as the same cork rest in the middle
of the water.*

*Then you shall take a deep glass, bowl, cup or other
vessel, and fill it with fair water, setting it in some
place where it may rest quiet, and out of the wind.
This done, cut the cork circumspectly by little and
little, until the wire with the cork be so fitted, that it
may remain under the superficies of the water, five*

or seven centimetres, both ends of the wire lying level
with the superficies of the water, without ascending or
descending, like to the beam of a balance being equally
poised at both ends.

In other words, after pushing the wire through the cork,
Norman carefully pared away the cork until it just floated
with the wire stuck through it.

He then took the cork out of the water and stroked it
with his loadstone, the north end with the north end of the
loadstone and the south likewise, and put it back in the water.

. . . and you shall see it presently turn itself upon its
own centre, showing the aforementioned declining
property. . .

This was a superb way of allowing the needle to rotate
in three dimensions, and therefore point toward the
strongest magnetic pull. Norman would not have been
able to do this mechanically; there would have been far too
much friction in the bearings.

Measuring latitude
He hoped that by measuring the angle of dip he would
be able to make an instrument that would measure
latitude directly, since it seemed reasonable to assume
that the angle of dip – or the amount of declination –
would steadily increase as you approached the North
Pole. Unfortunately it's not as simple as that, but he did
construct an elegant dip circle.

BIG OR SMALL: WHICH FALLS FASTER?

GRAVITY AND THE SCIENCE OF FALLING

1587
THE STUDY
RESEARCHER:
Galileo Galilei
SUBJECT AREA:
Gravity
CONCLUSION:
Objects fall at the
same rate regardless
of their mass.

Galileo Galilei, who lived most of his life in Pisa, Padua and Florence, was a towering figure in the early stages of experimental science. He had a clear and logical way of thinking about the world. He wrote 'Nature... does not do that by many things, which may be done by few.' This is similar to Occam's razor, which says that if you have to choose between competing hypotheses you should pick the one with fewest assumptions.

He also wrote 'Philosophy [i.e. Science] is written in... the language of mathematics, and its characters are triangles, circles and other geometric figures.'

He first came to fame as a medical student in 1581. Sitting one day in the superb cathedral or *duomo* in Pisa, and perhaps bored by a long sermon, he noticed the great bronze lamp swinging in the draft. It hung on a long chain from the top of the dome, and swung slowly from side to side. He timed it by feeling his pulse, and found to his surprise that it always took the same time for one swing, whether it swung a whole metre from side to side, or just a few centimetres.

Experiments with pendulums

Galileo went home and made some pendulums by tying weights on the end of strings, and investigated. He found that the width of the swing did not make much difference, nor did the weight on the end of the string. The only thing that mattered was the length of the string. To get a swing twice as slow he needed four times the length of string.

We now know that the time t in seconds is given by the formula $t = 2\pi\sqrt{l/g}$ where l is the length of the pendulum and g is the gravitational constant, 981 cm/s^2 (386.2 in/s^2).

Galileo realized that a pendulum would be an excellent way of regulating a mechanical clock, and he designed one, but did not get around to building it before he died in 1642; the first pendulum clock was made by the Dutch polymath Christiaan Huygens 15 years later.

Falling bodies

When he became professor of mathematics at Pisa in 1589, Galileo began to think about some of Aristotle's assertions, and in particular about falling bodies. Aristotle had said that the big things fall faster than small ones, and in particular that if you took two stones, one twice as heavy as the other, the heavy one would fall twice as fast.

Galileo wondered whether this was true, and decided to test it by experiment. Legend has it that he climbed up the famous leaning tower in the Piazza dei Miracoli, and dropped balls of various weights from the top, to see how fast they fell. This would have been a difficult task to execute, as even just dropping things simultaneously is not easy, and when they hit the ground they would be going so fast that it would be almost impossible to tell what happened when, let alone measure anything useful.

The inclined plane

What we know is that Galileo cut a groove along a wooden beam, polished it and lined it with parchment. Then he propped up one end of it and allowed a polished bronze ball to roll down from the top. By using this inclined plane, in effect he slowed down the fall, and he was then able to make careful measurements of how fast it went.

Timing was still difficult, since there were no precise clocks or watches; so at first he used his pulse, then a water clock and later he used sound. He rigged a series of tiny bells above the slope, so that the balls would touch them as they passed, and produce pings. By listening to the pings he was able to tell a good deal about the speed.

When Galileo placed the bells at equal intervals down the slope and rolled a ball down, he could hear the pings getting closer and closer together as the ball rolled; in other words it must be accelerating as it went down. By trying a variety of positions for the bells, he discovered that they pinged at equal intervals of time when they were placed at intervals of 1, 3, 5, 7 and 9 units down the slope, which meant at distances from the start of 1, 4, 9, 16, and 25 units. Thus he showed that a ball would roll 1 unit of distance in 1 second, 4 units in 2 seconds, 9 units in 3 seconds, 16 units in 4 seconds and 25 units in 5 seconds. The distance rolled was proportional to the square of the time.

Constant acceleration
He realized that the ball was accelerating at a constant rate, or as he put it, 'when starting from rest, acquires during equal time intervals equal increments of velocity'.

Galileo did not have the mathematics to work out the equations of motion; Newton would do that some decades later. However, Galileo did show that heavy balls and light balls rolled down the slope at the same speed. Aristotle was completely wrong.

1648

THE STUDY

RESEARCHER:
Blaise Pascal

SUBJECT AREA:
Meteorology

CONCLUSION:
Air pressure
decreases with
increasing altitude.

IS THE AIR THINNER ON MOUNTAIN TOPS?

THE PRESSURE OF THE ATMOSPHERE

Born in Rouen, France, Blaise Pascal was a child prodigy, and became a mathematician and physicist. He invented and built a calculating machine, and was a pioneer in pure mathematics and in the study of mathematical probability. It was his interest in the work of Galileo and Torricelli, however, that led to his discovery of the rise and fall of atmospheric pressure.

Galileo and Torricelli
Before Galileo died in 1642, he was told by the pump-makers of the Grand Duke of Tuscany that their pumps could not suck water up more than about 30 ft (10 m), which puzzled him, and he discussed it with his pupil Evangelista Torricelli, who was with him when he died.

Torricelli decided to investigate, using mercury, which is 14 times denser than water, and should therefore show the same effect at a height of less than 1 metre.

He made a glass tube about 1 metre (3 feet) long, sealed it at one end, and filled it with mercury. Then he put his finger over the open end and upended the tube in a bowl full of mercury. The level of mercury in the tube sank to about 76 centimetres (29 inches) above the surface of the liquid in the bowl.

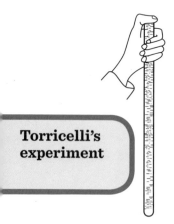

Torricelli's experiment

There was great controversy about the space at the top of the tube. Torricelli said it was a vacuum, but few believed him, since a vacuum was thought to be impossible. 'Nature abhors a vacuum,' as Aristotle said.

Torricelli may have noticed that the level of mercury in the tube seemed to go up and down with the weather; if so he had effectively invented the barometer. However, he died in 1647, so he never pursued this.

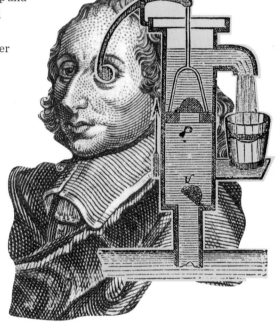

Pascal Experiments

Blaise Pascal was intrigued by Torricelli's work, and investigated the effect with a variety of liquids. He wondered what could be holding the liquid up in the tube; could it be the weight of the atmosphere pushing down on the liquid in the bowl? Would there be less force pushing down on top of a mountain, because there was less air above? He boldly predicted that at the top of a mountain the level of liquid would drop in the tube.

After considerable badgering, Pascal persuaded his brother-in-law Florin Périer to try the experiment by climbing 1,000 metres (3,900 feet) up the Puy de Dôme, an extinct volcano near Clermont-Ferrand in the centre of France. On 19 September 1648, Périer set off at 8 am from a monastery at the base, where he measured the height of the mercury, or quicksilver: 'I found the quicksilver stood at 26'and 3 1/2 lines above the quicksilver in the vessel [12 lines = 1 in]'.

Then, with some assistants, he lugged several 1.3 metre (four-foot-long) glass tubes and 7 kilograms (16 pounds) of quicksilver to the top of the mountain, where he 'found that the quicksilver reached a height of only 23' and 2 lines. . . I repeated the experiment five times with care. . .each at different points on the summit. . . found the same height of quicksilver. . . in each case.' In other words, the pressure was clearly lower at the top of the mountain.

Pascal's Principle

So Pascal had strong evidence to support his theory; it was indeed the weight of the atmosphere that holds up the column of mercury or water. In fact we now know that atmospheric pressure at sea level is around 15 pounds per square inch (psi), or just over 100 pascals (Pa), where 1 Pa is a pressure of 1 newton per square metre. An atmospheric pressure of 100 Pa is equivalent to 1 kilogram per square centimetre (kg/cm²); which means that there is about a kilogram pressing down on each of your fingernails. Luckily you have firm flesh underneath, which pushes back just as hard.

Pascal also showed that the pressure at the bottom of a column of liquid is proportional to its height, and allegedly attached a thin 10 metre (33-foot-long) vertical tube to the top of a barrel full of water, and poured water in at the top; the barrel burst.

He showed that in a closed vessel, the pressure is the same in all directions. This is known today as Pascal's Principle, and his discovery led to the invention of the syringe and the hydraulic press.

Pascal's barrel experiment

WHY ARE TYRES FILLED WITH AIR?

AIR PRESSURE AND THE POWER OF THE VACUUM

1660
THE STUDY
RESEARCHERS:
Robert Boyle and
Robert Hooke

SUBJECT AREA:
Pneumatics

CONCLUSION:
The volume of
affixed mass of gas is
inversely proportional
to its pressure.

Robert Boyle, the seventh son of the Great Earl of Cork, was born at Lismore Castle on the south coast of Ireland, on 25 January 1627. As a teenager he travelled to Europe with a French tutor, and visited Galileo before he died in 1642. Boyle came home determined to be a scientist. He joined the 'invisible college', which met in London or Oxford to cultivate the 'new philosophy', and this eventually became the Royal Society of London for improving natural knowledge – now the Royal Society.

The Magdeburg hemispheres

In 1654, in Magdeburg in what is now Germany, the Burgomeister (mayor) Otto von Guericke, who was an enthusiastic scientist, built an air pump. He used this to demonstrate the power of the vacuum, or more accurately the power of atmospheric pressure. In 1657 he used his machine to pump the air out of two 30 centimetre (12 inch) brass hemispheres, so that they were held together by air pressure. Two teams of horses were unable to pull them apart, until he let the air back in.

Meanwhile, Boyle had inherited some land in Ireland, and with it a fortune, and settled in Oxford. Knowing about the

> **Magdeburg hemispheres**

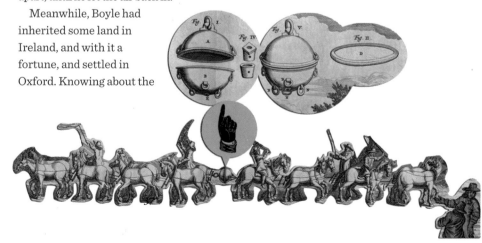

work of Torricelli and Pascal (see pp. 34–6), and having heard about the Magdeburg hemispheres, he hired Robert Hooke to build an air pump, with which they did a series of experiments that he reported in his 1660 book *New Experiments Physico-Mechanical: Touching the Spring of the Air and their Effects.*

Experiments with the air pump

Boyle and Hooke were able to pump almost all the air out of a large glass bell jar, so as to create inside what was almost a complete vacuum; the pressure was probably less than one-tenth of normal atmospheric pressure. They would set up an experiment inside, and then pump the air out, and this is what they found:

- A burning candle went out; so air was necessary to support a flame.
- A bell ringing inside could not be heard from outside; so air was necessary for transmission of sound.
- Red-hot iron still glowed; so air was not necessary to transmit light.
- A bird and a cat inside died; so air was necessary for life.

Experiments with the J-tube

In Boyle and Hooke's experiments a fixed amount of air is trapped in the closed end of the tube by mercury in the bottom. The pressure on this air can be reduced by putting the whole thing inside the bell jar and pumping the air out of the bell jar, or it can be increased by putting more mercury into the open end. Boyle and Hooke noticed that decreasing the pressure increased the volume, and increasing the pressure decreased the volume, but Boyle did not write up the details at this point.

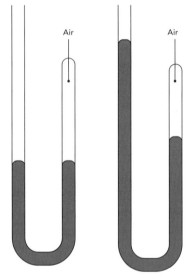

Air

Air

J-tubes

Meanwhile in Lancashire, Richard Towneley of Towneley Hall did his own experiments with physician Henry Power. They also used a J-tube, and on 27 April 1661, took a sample of 'valley ayr' in the tube 300 metres (1,000 feet) up Pendle Hill. At the top, where the atmospheric pressure was lower, they noted that the volume had increased. Then they put a sample of 'mountain ayr' in the tube, and went down the hill again; at the bottom the volume had decreased.

Towneley discussed this experiment with Boyle that winter, and suggested that there might be an inverse relationship between volume and pressure. Boyle then did his own quantitative experiments, and was meticulous about recording his observations; the answer was that the volume of affixed mass of gas is inversely proportional to its pressure. This became known as Boyle's Law, but Hooke, Boyle himself, and Isaac Newton called it 'Mr Towneley's hypothesis'.

The spring of the air
Boyle imagined that the individual particles of air were like little coils of wool, which would be compressed like springs by increased pressure, but would bounce back again when the pressure was released. That is why he talked about the spring of the air, and why we use pneumatic tyres on cars and bicycles; the spring of the air irons out the roughness of the road. According to one account, Torricelli did not notice the fact that a column of mercury does not always reach exactly the same height in his tubes. Boyle and Hooke did notice, and wondered whether the variation was due to the tides. When they checked, they found no correlation with high or low tides; instead the height of the mercury was greater in fine weather, and sank in poor weather, especially in storms. Therefore the barometer was really invented by Boyle and Hooke, rather than by Torricelli – though he did show them the way.

1672

THE STUDY

RESEARCHER:
Isaac Newton

SUBJECT AREA:
Optics

CONCLUSION:
White light is a
mixture of all the
colours of the
rainbow.

IS 'WHITE' A COLOUR?

UNRAVELLING THE NATURE
OF WHITE LIGHT

Isaac Newton was a sickly boy. When he was born, on Christmas Eve 1642, he was so small and feeble that they did not expect him to last the night. When he was three years old his father died, and his mother went off with a rich clergyman, leaving him in the not-too-tender care of her parents. He grew up lonely and introspective, but acquired a remarkable ability for focusing on a range of problems, from the colours of the rainbow to the orbits of the moon and the planets, which made him perhaps the greatest scientist of all time.

In the late 1660s he designed and built a reflecting telescope – one of the first to be made. Later he made a second. He was by then the Lucasian Professor of mathematics at the University of Cambridge, and spoke about his telescope in his lectures. When they saw it, the authorities at the Royal Society elected him a Fellow, and asked him what else he had been doing. He responded with a letter, dated 6 February 1672, describing in detail his experiments with prisms.

The spectrum

'Having darkened my chamber, and made a small hole in my window-shuts [shutters], to let in a convenient quantity of the Sun's light, I placed my prisme at his entrance, that it might be thereby refracted to the opposite wall.'

This formed a spectrum five times as long as it was broad, which surprised Newton, and he tried adjusting the prism so that it was outside the shutters, and so that the light passed through a thicker part of it, and making a bigger hole. None of these things made any difference; so he concluded this must be the effect of refraction on the sunlight.

With care he measured the distance across the room, and calculated the angles of refraction, and showed that the blue light is refracted more than the red. Newton claimed he could see seven colours in the spectrum – red, orange, yellow, green, blue, indigo and violet. Most people stop at blue, although there are different shades of blue. Newton's eyes may have been especially sensitive to the far blue, or he may have decided that there should be seven colours, since seven was a number of mystical importance to him.

Then I began to suspect, whether the rays... did not move in curve lines, and according to their more or less curvity tend to divers parts of the wall. And it increased my suspition [sic], *when I remembered that I had often seen a Tennis ball, struck with an oblique Racket, describe such a curve line.*

As the tennis ball spun, one side would experience more air resistance than the other, and he guessed that the same might be true for particles of light – for he believed

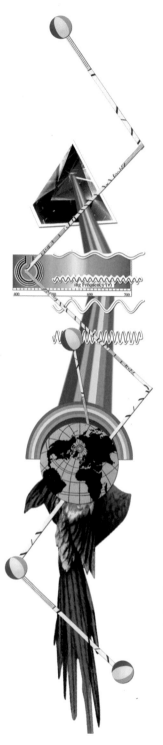

that light was made of particles (or 'globular bodies'). But he showed the rays actually went in straight lines.

Then Newton went on to what he called the *Experimentum crucis*. He made a small hole in a board and placed it between the prism and the wall, so that he could isolate one colour at a time. When that colour – say green – passed through the hole he passed it through a second prism, and found that it was refracted again to give a splash of green light on the wall, and the angle of refraction was the same this second time. He also showed that it stayed green; he could neither change its colour nor separate it into further colours.

What is white light?

He concluded that sunlight 'consists of Rays differently refrangible, which... were, according to their degrees of refrangibility, transmitted toward divers parts of the wall.' In other words, white sunlight comprises a mixture of all the colours, which are separated by the prism because they are refracted through different angles. 'Why the Colours of the Rainbow appear in falling drops of Rain, is also from hence evident.'

Finally, he used a lens (or a second prism) to bring all the colours back together, and make white light. In a four-paragraph aside, he described how he realized that a reflecting telescope would be free of the coloured fringes that always appeared with conventional lens-based telescopes, and how he made one and used it to observe the moons of Jupiter, and Venus as a crescent.

Newton went on to say that 'the Colours of all natural Bodies have no other origin but this, that they are variously qualified to reflect one sort of light in greater plenty than another.' In his darkened room he put various objects in his spectrum of colours, and showed that anything can be made any colour, but 'they are most brisk and vivid in the light of their own daylight-colour'.

DOES LIGHT TRAVEL AT A FINITE SPEED?

SEARCHING FOR THE SPEED OF LIGHT

1676

THE STUDY

RESEARCHER:
Ole Rømer

SUBJECT AREA:
Optics

CONCLUSION:
Rømer measured the speed of light to be about 220,000 kilometres per second.

Light travels extremely fast; indeed for hundreds of years people thought it was instantaneous; that going from A to B took zero time.

Galileo was not convinced, and tried to measure the speed of light in 1667. He stood on a mountain with a lantern, while a friend stood on another, a mile away, with another lantern. Galileo raised the shutter on his lantern, and as soon as the other man saw the light he opened his shutter too. Galileo saw the light coming back within a quarter of a second, which was probably the reaction time of his friend – the time it took him to start moving when he saw the light. Galileo concluded that the speed of light is too high to be measured in such a way.

In his late twenties, Danish scientist Ole Christensen Rømer was invited from Copenhagen to Paris, where he became Royal Mathematician and tutor to the son of Louis XIV. He did a lot of research at the Royal Observatory, where the director was the Italian astronomer Giovanni Domenico Cassini, who had discovered the gap in Saturn's rings, which is still called the Cassini division.

The moons of Jupiter

Cassini was searching for a solution to the problem of finding longitude at sea. One possibility, suggested by Galileo, was by observing the four biggest moons of Jupiter, discovered in 1610. The moons orbit Jupiter in a regular way, and in particular Io, the closest of the

Galilean moons and about the same size as our moon, orbits in slightly less than two days.

The moons can be seen on one side of Jupiter, and then disappear into the shadow behind the giant planet before emerging again into the sunlight. If a sailor could measure the time of appearance of Io into the sunlight, and compare that time with printed tables, he should be able to read off his longitude because the time varies slightly according to where on Earth you are.

There are various problems. For one thing, if you are waiting for Io to appear you need continuous observation for many minutes, which is difficult or impossible if the sky is cloudy. Also, observing the moons from a steady position on dry land is easy – you need only binoculars or a small telescope – but doing so from a moving ship is close to impossible. So this was never really going to be a practicable solution to the problem of longitude.

Nevertheless, the astronomers at the Paris Royal Observatory had gathered masses of data on the emergence of Io, and Cassini had published tables predicting when it would be visible from various places on Earth.

Spotting the anomaly

Rømer realized that there was magic in that data. The tables were a bit suspect, and needed occasional corrections, but there was a recurring discrepancy according to the relative positions of Earth and Jupiter.

For several months Jupiter is not visible from Earth, because it is either directly behind the sun, or close enough for the sun's light so that it is too bright to allow observation. When Jupiter first becomes visible, however, it is a long way from Earth (distance A, on the opposite diagram). As Earth moves around in its orbit, it gets steadily closer to Jupiter until it reaches the nearest point (distance B, on the opposite diagram), then it moves further away.

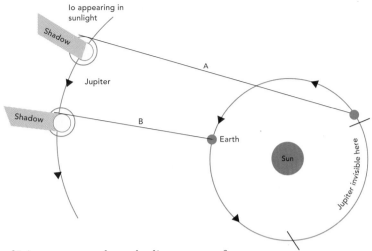

The times of Io's appearance showed a discrepancy of 11 minutes between the time when Jupiter was only just visible past the sun, and the time of nearest approach; in other words between when Jupiter was furthest away and closest. This had to mean that light took 11 minutes to cover the difference in distance: A minus B.

RøMer did not have an accurate value for the distance from Earth to the sun, but using the best estimate available, he was able to calculate how far light must have travelled in those 11 minutes, and so worked out that the speed of light must be 214,000 kilometres per second (133,000 miles per second). This was about 25 per cent short of the speed we accept today (about 299,792,458 kilometres per second or 186,000 miles per second), but it was the first measurement, and astonishingly close.

In September 1676, Rømer predicted that the appearance of Io on 9 November would be ten minutes later than the time shown in the table, and he was exactly right. In spite of this, Cassini refused to accept the reasoning, and Rømer never published his results officially. When he visited England, however, he found that Newton and Edmond Halley agreed with him and gave him their support. On his return to Copenhagen he became Director of the Royal Observatory.

Finding longitude in space

1687

THE STUDY

RESEARCHER:
Isaac Newton

SUBJECT AREA:
Mechanics

CONCLUSION:
Objects move in
straight lines at
constant speeds
unless forces act
on them.

IS THE 'FALLING APPLE' STORY TRUE?

THE LAWS OF MOTION

Newton was born in Lincolnshire, England, and that was where he went when Cambridge University was closed because of the plague in 1665. He stayed there for 18 months, and probably did most of his finest scientific work in that time – a loner with time to think.

According to legend, he saw an apple fall from a tree – there is a very old apple tree in front of the house – and thought that something must pull the apple down from the tree; so a pulling force extends from the Earth upwards, at least to the top of the apple tree. Could it extend as far as the moon? If so it would affect the moon's orbit. Could it actually cause the moon's orbit?

Supposedly he grabbed his mother's title deeds and began doing calculations on the back. He realized that the power of attraction would get less as the height of the object increased, and he guessed that it would decrease with the square of the distance between the object and the centre of Earth. The results of his calculations, he said, 'seemed to answer, pretty nearly'. He further guessed that this sort of attraction would be true for other orbital motions, and he called it 'universal gravitation'.

We hear nothing of this for nearly 20 years. Then three friends, Edmond Halley, Robert Hooke and Christopher Wren, meeting as usual in a coffee shop in London, got to arguing about the path of a comet as it came towards the sun. Hooke said he could calculate the mathematics, but in practice failed to do so.

A visit to Cambridge

Halley was one of Newton's few friends, and in 1684 visited him in Cambridge. He asked Newton what would be the trajectory of a comet, assuming an inverse-square law of attraction. Newton at once said it would be an ellipse. He knew, he said, because he had calculated it earlier – but then he could not find the calculation among his papers. So he promised to do it again and send it to Halley.

In November that year Newton produced the paper, 'De motu corporum in gyrum' (On the motion of bodies in an orbit), explaining the effects of the inverse-square law, and then in 1687 his monumental work *Philosophiae Naturalis Principia Mathematica* (The mathematical principles of natural philosophy), generally just called the *Principia*.

In this difficult book, written in Latin, Newton explained not only the inverse-square law and his concept of universal gravitation, but what have become known as Newton's laws of motion, although the first two were already well known. *Principia* lays down the foundations of classical mechanics.

The apple story

William Stukeley was an antiquarian – a historian and archaeologist who pioneered the investigation of Stonehenge. He was also a friend of Isaac Newton and his first biographer. Stukeley wrote vividly (and proudly) about the events of 15 April 1726:

I visited Sir Isaac Newton... & spent the whole day with him. After dinner, being a fine day, we sat in the garden, under the apple trees, and drank tea there. He told me among other discourse it was in such a situation, that he first took the notion of the gravitation of matter:

from an apple dropping off a tree. Why should this
apple, always and invariably fall to the earth, in a
perpendicular line; why should it not fall upwards,
sideways or obliquely?

Such questions, said Stukeley, 'revolved in his mind'
and 'thence he began to consider, & discover, the mode, &
laws of this universal power in matter & apply them to the
motion of the heavenly bodys, to the cohesion of matter; &
to unfold the true philosophy of the universe.'

An assistant of Newton's, John Conduitt, also wrote
about the apple in his 1727 biography of Newton: 'In the
year 1666 he retired again from Cambridge to his mother
in Lincolnshire. Whilst he was pensively meandering in a
garden it came into his thought that the power of gravity
(which brought an apple from a tree to the ground) was not
limited to a certain distance from earth, but that this power
must extend much further than was usually thought.'

So Newton told at least two people about the apple, but
this was 60 years after he claimed it had happened, and it
is possible that he simply made up the story.

Why should he make it up?

Until 1682 Newton's letters show that he believed the
planets are swept around the sun in a great vortex, like
water going down a plug hole, as originally suggested
by Descartes. But Halley's comet of 1682 upset that
idea, because it had a retrograde orbit – that is, it moved
backwards: in the opposite direction to all the planets.

But Hooke had written about gravity in 1674, and
was getting close to solving the mathematical problem.
Newton could never admit that Hooke had beaten him to
anything; so it is just possible that Newton made up the
apple story, long after the event, in order to prove that he
had sorted it all out in 1666, well before Hooke.

IS ICE HOT. . . ?

THE DISCOVERY OF HIDDEN HEAT

1760
THE STUDY

RESEARCHER:
Joseph Black

SUBJECT AREA:
Thermodynamics

CONCLUSION:
Heat is needed to turn ice into water, and water into steam.

In spite of his Scottish heritage, Joseph Black was born in the south of France, because his father was a wine merchant, and his family owned a town house in Bordeaux and a vineyard with a farmhouse nearby.

He must have had a nasty shock going to school in cold Belfast, and then to Glasgow University, where he took to science and medicine. Working for his doctorate in the early 1750s Black was the first to isolate a pure gas, 'fixed air', which we call carbon dioxide.

Melting snow

The winters of 1755 and 1756 were particularly cold, and when Black became a professor at Glasgow in 1757, he began to wonder about the melting of ice and snow, and said in his lectures:

If we attend to the manner in which ice and snow melt... however cold they might be at first, they are soon heated up to the melting point, or begin soon at their surface to be changed into water. And if... the complete change from them into water required only the further addition of a very small quantity of heat, the mass, though of a considerable size, ought all to be melted in a very few minutes.... Were this really the case... the torrents and inundation would be incomparably more irresistible and dreadful.

Knowing that snow and ice can persist for weeks or even months Black concluded that ice and snow do not melt easily. But why?

He observed that 'Even without the help of thermometers we can perceive a tendency of heat to diffuse itself from any hotter body to the cooler around, until it is distributed among them ... The heat is thus brought into a state of equilibrium.'

Using his thermometer, he showed that when he added 500 grams (1 pound) of hot water to 500 grams of cold water he finished with 900 grams at a temperature midway between the two.

Next, he set up an ice experiment. He filled two identical flasks with water and cooled flask A almost to freezing point, 0°C (32°F), and the second, B, to just below 0°C, so that the water inside turned to ice. He hung the flasks side-by-side in a still room, and waited for them to warm to room temperature. The water in flask A got there in half an hour, but the water in flask B took more than ten hours. The ice had clearly needed heat in order to turn it into water, before the temperature could start going up.

Latent heat

Black called the heat that he could feel with his hand – which can be measured as temperature – 'sensible' heat, while the extra heat needed to melt the ice he called 'latent' heat, meaning hidden heat.

To test his theory he took two more flasks, and filled C with water, and D with a mixture of water and alcohol. He put a thermometer in each and left them outside on a cold night. The temperature in both flasks dropped gradually to 0°C . Then in flask C it stayed at 0°C while ice formed around the thermometer. In flask D the temperature went on falling, because the water-alcohol mixture did not freeze.

Boiling water

He then investigated the boiling of water in the same way – and you can do this yourself, if you have a suitable thermometer that covers roughly the range 20–110°C (65–220°F).

Put a thermometer in a pan of water and place it on a hot stove. The temperature gradually rises until it reaches 100°C (212°F). Then the water starts boiling, and the temperature does not go up any more. Put more heat in, on a hotter stove, and the water boils faster, but the temperature stays the same.

To boil water takes heat. Heat is needed to give each molecule of water enough energy to escape from the liquid and turn into steam. This also is latent heat – the latent heat of evaporation.

James Watt and his separate condenser

Black's discovery of latent heat was almost certainly the spur that in 1765 pushed his friend James Watt to invent the separate condenser, and transform the efficiency of steam engines.

In 1766 Black moved to Edinburgh University, where many of his students were the sons of whisky distillers. They asked him why so much fuel was needed to heat their stills, making whisky so expensive. His answer was simple: latent heat. They had to supply energy to turn the liquid into vapour, before cooling it down again into whisky.

CHAPTER 3: Wider fields: 1761-1850

During the 18th century scientists took on broader challenges. Measuring the mass of the Earth must have seemed like an impossible task in Newton's time, but then two different methods surfaced, along with one reluctant astronomer, and one reclusive genius, to carry them out.

The invention of the electric battery changed both science and the world forever, since it founded several new sciences and led to all the gadgets we have today. Patient brewer James Joule spent years unravelling

the mechanical equivalent of heat, despite sceptical
receptions from other scientists.

   Further arguments cropped up about the nature
and behaviour of light. When the connection between
electricity and magnetism was spotted, Michael Faraday
and others jumped in, investigated and acted on their
findings, with the building of the electric motor, the
transformer, the electromagnet and the dynamo.

1774

THE STUDY

RESEARCHER:
Nevile Maskelyne

SUBJECT AREA:
Gravity

CONCLUSION:
Earth is not
hollow, but has
a metallic core.

CAN YOU WEIGH
THE WORLD?

A HEROIC EXPERIMENT USING MOUNTAINS

In his famous 1687 book *Principia,* Isaac Newton
mentioned the fact that a plumb bob would always hang
vertically down towards the centre of Earth, unless
there was a mountain nearby, in which case the bob
would be pulled a little sideways by the
gravitational effect of the mass of the
mountain. He called this 'the attraction of
mountains', but thought that in practice
the effect would be too small to measure.

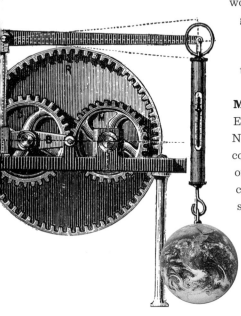

Measuring the attraction of mountains
Eighty years later the Astronomer Royal,
Nevile Maskelyne, realized that if the effect
could be measured, it might provide a way
of measuring the mass of Earth. If someone
could hang a plumb bob beside a mountain,
see how much it was pulled sideways and
estimate the mass of the mountain, then
he could work out the mass of Earth.
This was important, for it would allow
him to calculate the masses of the
moon, the sun and the other planets
too. In 1772 he sent a proposal to the
Royal Society.

The Royal Society approved of the idea, and
dispatched surveyor Charles Mason around Scotland
on horseback to find a suitable mountain. He returned
after a long summer's touring, and reported that the best

mountain he could find was Schiehallion, 72 kilometres
(45 miles) north of Perth.

Who should do it?

Maskelyne said he was too busy to undertake the
experiment, and anyway he was
the Astronomer Royal – which
meant he would have to get
permission from the King.
Unfortunately the King was all for
it, and gave him temporary leave
of his duties; so the reluctant
Maskelyne left his comfortable
lodgings in Greenwich and
headed north into the Highlands.

In the mountains

Schiehallion, 1,083 metres (3,543
feet) high, is a long, narrow
mountain, lying roughly East-West.
Maskelyne set up camp halfway up
the middle of the South side. He had a bothy (a small hut),
a large tent, a precision pendulum clock and a three-metre
(10 foot) telescope borrowed from the Royal Society. He
planned to work out his precise position by observing the
stars overhead, using a plumb bob to determine the 'vertical'.
Unfortunately there was so much mist and rain that he
could do no observing at all for two months, and it took him
another month to work out exactly where he was.

Then he moved round to the north side of the mountain
– the move took an entire week – and did the same thing
again. Meanwhile a team of surveyors, housed in rough
tents, tramped around the mountain with chains (for
measuring length), barometers (for measuring altitude),
theodolites (for measuring angles) and other surveying
equipment. They recorded thousands of angles and

altitudes at different points, and worked out the distance between Maskelyne's two camps.

Maskelyne, using the stars above and the plumb bob, worked out the apparent positions of both his camps, and calculated how far apart they were. His measurement and the surveyors' measurement differed by just 436 metres (1,430 feet), because he had been using verticals distorted by the attraction of the mountain.

The discrepancy was less than he had expected, which meant that the average density of Earth was much greater than that of the mountain. This laid to rest one earlier theory, that Earth was hollow in the middle, like a tennis ball. On the contrary, said Maskelyne, it must have a metallic core. Now all he had to do was work out the mass of the mountain. He could estimate its density – that is its mass per unit volume – but he needed to find the volume of this awkward shape.

Calculating the volume

To do this he enlisted the help of a mathematician friend. Charles Hutton realized that he could use all the height measurements from the surveyors to work out the three-dimensional shape of the mountain. He wrote in his report that he had joined together with faint pencil lines all the points of equal height, which immediately gave him a good idea of the shape. In other words he invented what we know as contour lines.

Knowing the volume of the mountain, Maskelyne and Hutton could work out its mass, and therefore the mass of the Earth, which came to about five thousand million million million tonnes. In the previous century, Newton had estimated the mass of the Earth at six thousand million million million tonnes, which in the end turned out to be more accurate. Nevertheless Maskelyne's heroic experiment was the first attempt at the measurement of the mass of Earth.

CAN YOU WEIGH THE WORLD (WITHOUT USING A MOUNTAIN)?

AN ALTERNATIVE METHOD TO MEASURE THE EARTH'S MASS

1798

THE STUDY

RESEARCHER:
Henry Cavendish

SUBJECT AREA:
Earth science

CONCLUSION:
The mass of Earth is six thousand million million million tonnes.

John Michell became Professor of geology at Cambridge, and lectured on arithmetic, geometry, theology, philosophy, Hebrew and Greek, but retired at the age of 37 to take a lucrative job as Rector of St Michael and All Angels at Thornhill in Yorkshire, probably in order to have more money and time for scientific research. He was the first person to propose the idea of a black hole, in a letter to the Royal Society in 1784. He also designed and built a piece of apparatus for measuring the mass of Earth, but never got around to actually doing the experiment. When he died in 1793 he left it to his friend Henry Cavendish.

Henry Cavendish was an extraordinary man. Today we might say he had Asperger's syndrome. The grandson of two dukes, he was enormously wealthy, and built his own laboratory in his house in Clapham Common in London, England. He was said to be the richest of all learned men, and very likely also the most learned of the rich.

The silent genius

Cavendish always wore a crumpled purple suit and a three-cornered black hat. Extremely shy, he avoided people. When he did speak it was with a high squeaky hesitant voice, and as a result he almost never spoke. One of his colleagues said that he probably spoke fewer words in his life than a Trappist monk. He attended meetings of the Royal Society, silently.

In 1766 he isolated hydrogen gas – the second pure gas ever isolated – found it was very light and very flammable, and by exploding mixtures with air showed that the only product was water, and that the formula of water was H_2O, although he told James Watt about it and Watt published the information in 1783.

Weighing the world

Cavendish set up Michell's apparatus, and prepared to measure the mass of Earth, to test Maskelyne's findings of 20 years earlier. The experiment was really a simpler and more refined version of the attraction of mountains, but using lead balls instead of the mountains.

On a long, thin wire hung a horizontal wooden rod 1.8 metres (6 foot) long. On each end was a lead ball with diameter 5 centimetres (2 inches) and mass 0.73 kilograms (1.61 lb pounds). On the anticlockwise side of each of these balls (as seen from above), 23 centimetres (9 inches) away, was a big lead ball with 30 centimetre (12 inch) diameter and mass 159 kilograms (350 lb).

The idea was that the small balls would feel a gravitational attraction from the big balls, and so be pulled towards them by a tiny amount; the beam would rotate until the effort of twisting the fine suspending wire just balanced the attractive force between the balls. Cavendish knew the weight of the small balls – that is their force of attraction towards the Earth. If he could measure the force of attraction towards the large balls then he would know the ratio of the mass of the Earth to the mass of the large balls.

Sensitive apparatus

The apparatus was left for several hours to settle down; this was such a delicate experiment that it would have been spoilt by a tiny draught or a slight temperature change. So Cavendish isolated it in a separate room, adjusted it with external controls and made observations through a window using a telescope.

Once it had settled down, and the small balls were still, Cavendish recorded their position, and then moved the large balls round to the other side of the small ones, so that the small ones would be pulled the other way. Once the small balls had settled again he found they had moved just 4.1 millimetres (0.16 inches). Cavendish measured this with amazing precision, and so was able to calculate the force that had pulled them sideways.

This force was extraordinarily small – about 15 nanograms, or the same as the weight of a small grain of sand – but that was enough. Cavendish was exceptionally careful to eliminate all possibilities of error, and his data led to an average density for Earth of 5.4 times the density of water, and a mass of the Earth close to the presently accepted value of 5.97 thousand million million million tonnes (5.97×10^{24} kg).

This experiment is now frequently done by physics students, and although the idea and the apparatus came from John Michell, it is called 'the Cavendish experiment', after the man who conducted it.

The Cavendish experiment

1799

THE STUDY

RESEARCHER:
Alessandro Volta

SUBJECT AREA:
Electricity

CONCLUSION:
This was the
foundation for several
new branches
of science.

BATTERIES NOT INCLUDED?

MAKING THE FIRST ELECTRIC BATTERY

Static electricity was well known to the ancients – the word 'electron' comes from the same Greek word, meaning amber, because the ancient Greeks knew that they could make static electricity by rubbing a piece of amber with a cloth. Benjamin Franklin showed by flying a kite into a thundercloud that lightning is just a form of electricity, but no one has yet managed to tame lightning. In order to study the properties of 'the electric fluid', scientists needed to be able to make it in reliable small doses, and continuously. The first step was taken at the University of Bologna in 1780, when Italian scientist Luigi Galvani was investigating the idea that animals are driven by electricity. He was busily dissecting frogs when he saw a frog's leg twitch. It was lying on the bench close to a static-electricity generating machine. The leg twitched again when he hung it out to dry on a brass hook that came into contact with a piece of iron. These observations supported Galvani's theory that the electricity was coming from the frog, even though it was long dead. Meanwhile, Alessandro Volta was Professor of Natural Philosophy (physics) at the University of Pavia. He was intrigued by Galvani's account of the twitching leg, but did not believe in animal electricity. He reckoned the electricity came from the touching of the two different metals. He published his ideas in 1793 and 1794, and launched an investigation.

Dissimilar metals

Volta took a piece of zinc and a piece of silver, like coins, put them together, and when he touched them on his tongue, he could feel a tingle. Then he had the bright idea of multiplying the effect by making a whole lot of these junctions, and connecting them in a row.

Just connecting zinc-silver-zinc-silver would be no good, for the effect would simply be cancelled at every other junction. What he needed was to separate each junction with something that would conduct electricity but was not a metal, in other words a non-metallic conductor. He used cardboard soaked in salty water. So he made a pile of zinc-silver-cardboard-zinc-silver-cardboard-zinc, and so on. This came to be called a 'pile' or 'battery'. His first crude device probably generated a few volts of electricity, but it was enough to give him a real shock, and enough to make a spark when he connected the ends together with a piece of wire.

Volta made this discovery in 1799, and he described his experiments in a long letter dated 20 March 1800, to Sir Joseph Banks, the President of the Royal Society in England:

I provide a few dozens of small round plates or discs of.
.. silver, an inch in diameter more or less, and an equal
number of plates of... zinc, nearly of the same size. I
prepare also... circular pieces of pasteboard
... capable of imbibing and retaining a great deal of...
salt water.

Volta's letter describes how to connect the bottom of the column to a bowl of water with a thick wire. 'A person who now puts one hand into this water, and with a piece of metal touches the summit of the column, will experience shocks and a pricking pain as high as the wrist of the hand plunged in the water, and even sometimes as high as the elbow. . .'

He also says 'The hearing will be strongly affected by introducing into the ears two probes, the opposite ends of which are connected with the two ends of the apparatus.'

Looking back, it seems curious that all Volta did was make sparks and give himself electric shocks, but this was only the beginning. Banks read out this letter, and immediately other scientists began to make their own batteries, which allowed them to make a continuous current of electricity; this had never been possible before.

For one thing, they could investigate the properties of materials, and find out about conductors and insulators. They could investigate the properties of electricity itself, and find out about potential (called volts, after Volta), current (amperes), resistance (ohms) and so on.

Electricity in chemistry

At the Royal Institution in London, Humphry Davy made an enormous battery, and used it to do some spectacular chemistry. He reckoned that putting dissimilar metals together must cause some chemistry to happen, which produces the electricity. Therefore he reasoned that he could use electricity to make chemistry happen, and he managed for the first time to isolate the metals sodium and potassium.

Today most of the things we use seem to rely on electricity. Volta's experiments were arguably some of the most creative in the history of science.

WHAT HAPPENS WHEN LIGHT UNRAVELS?

YOUNG'S DOUBLE-SLIT EXPERIMENT

1803
THE STUDY

RESEARCHER:
Thomas Young

SUBJECT AREA:
Optics

CONCLUSION:
Light travels in waves
– or does it?

In his famous paper of 1672, and in his 1704 book on *Opticks*, Isaac Newton wrote about 'rays' of light, but in the course of the book he gradually tends towards the idea that light comes in particles, or 'corpuscles'. This became known as Newton's corpuscular theory of light. The Dutch polymath Christiaan Huygens disagreed. He thought that light was made of waves, and the argument was unresolved for a hundred years.

Particles or waves?
Thomas Young was another polymath, who achieved much in various fields, and in the early 1800s published a series of papers describing refraction of light. He comes down on the side of light being a wave motion, which is supported by his results. He knew that when two slightly different notes sounded, he could hear beats, because of interference of the sound waves, and he reckoned that if light was indeed a wave motion, he should be able to get interference with the light waves.

Following Newton, he made a small hole in the window shutter, covered it with a piece of black paper and made a tiny hole in that with a needle. Then he set up a mirror so that the incoming beam of sunlight went straight across the room to the opposite wall:

I brought into the sunbeam a slip of card, about one thirtieth of an inch in breadth, and observed its shadow, either on the wall or other cards held at different distances. Besides the fringes of colours on each side of the shadow, the shadow itself was divided by similar parallel fringes.

The best-known of many such experiments is the one in which a beam of light shines through two parallel slits in a card, and on to a screen. This is generally known as 'Young's experiment', although there is no evidence that Young actually used two slits.

Interference pattern

If the beam comprises particles of light this should produce just two bright lines of light on the screen behind. What actually appears is an array of fringes.

Each slit behaves as a new source of light, and sends light out in a new set of waves. When the crest of a wave from slit A hits the screen at the same place as a crest from slit B, the result is a bright band. When a crest from A meets a trough from B they cancel out, and there is a dark band on the screen.

The result is that right across the screen there are light and dark bands, which can only come from refraction and interference of the two beams, and this means that light comes in waves. Even though Young did careful and thorough experiments, and put forward sound reasoning, many scientists were reluctant to believe him, for how

Particle pattern

Wave pattern

could the great Isaac Newton possibly be wrong? Not until light was shown to travel more slowly under water, 50 years later, was Young finally vindicated.

The distance between successive bright bands on the screen is a function of the wavelength of the light, and therefore varies according to the colour of the light.

Light is now thought to travel in packets of waves, called photons. One startling fact, which Young could not have known, is what happens in extremely low light. Photons can be arranged to arrive at the double slit one at a time. Each photon can go through only one slit, and so it should travel straight on, since there is no other photon to interfere with. The single photons certainly arrive in single spots on the screen. If the screen is a camera sensor, and the image can be left to build up over a long period, you might expect the particle pattern to emerge.

Wrong. Once again a pattern of bands is formed. Suddenly we have stepped into the weird world of quantum mechanics. According to quantum mechanics any one photon may not necessarily be all in one place. Suppose it has a 30 per cent chance of going through slit A and a 70 per cent chance of going through slit B, then quantum mechanics says that it can go through both slits, and interfere with itself.

In other words the photons are behaving as both particles and waves; this is called wave-particle duality. Those who believed in the particle theory of light were not wrong after all.

In 1961 the same phenomenon was shown for electrons; they must be particles, since they have mass, but they also behave as waves. In 1974 single electrons were shown to form interference patterns.

Richard Feynman called it:

a phenomenon which is impossible. . . to explain in any classical way, and which has in it the heart of quantum mechanics.

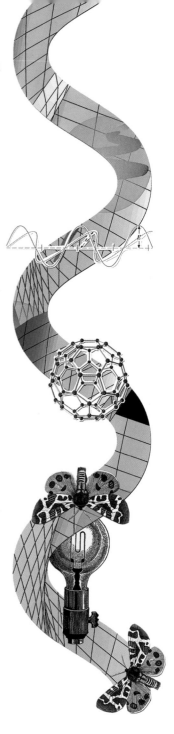

1820

THE STUDY

RESEARCHERS:
Hans Christian Ørsted
and Michael Faraday

SUBJECT AREA:
Electromagnetism

CONCLUSION:
Electricity and
magnetism can
be made to work
together.

CAN MAGNETS CREATE ELECTRICITY?

THE DISCOVERY OF ELECTROMAGNETISM

Electric batteries had been around for 20 years, and all sorts of scientists had used them for experiments, but no one had systematically looked for a link between electric currents and magnetic fields.

On 21 April 1820, Hans Christian Ørsted, professor of physics at the University of Copenhagen, was preparing a lecture for his students, when he noticed a twitch of the needle of a compass lying on the bench. It happened when he switched on the current from an electric battery, and again when he switched it off.

This was not entirely a chance discovery, since he had been looking for a connection between electricity and magnetism. He investigated, and found that an electric current flowing through a wire produces a circular magnetic field around the wire, like a sleeve. Three months later he published his results in privately circulated pamphlet.

Paris and London

At the French Academy of Sciences François Arago and André-Marie Ampère heard of Ørsted's work, and jumped into the fray. Ampère showed that two parallel wires carrying currents repel each other if the currents are flowing in the same direction, and attract each other if the currents are opposite. He went on to develop a mathematical theory to explain this: Ampère's Law says that the force between two such wires is proportional to the strength of the currents.

Ørsted's news went also to London's Royal Institution, where Humphry Davy and William Hyde Wollaston set out to make an electric motor, but failed. Davy's assistant at the time was Michael Faraday. He heard Davy and Wollaston talking about their electric motor, and he then went away and thought about the problem on his own.

During the course of a week in early September 1821 Faraday did a series of experiments investigating the attraction and repulsion of compass needles when close to a wire carrying a current. He drew diagrams of the observations, and finished up with a diagram of a wire moving round the end of the compass – that is, a magnet. This led him to make a toy motor.

The first motors

The first electric motors were simple devices. They each had a pool of mercury in a glass cup. On the left, a current was passed through a rigid brass rod which came down from above and just reached the pool of mercury, while a magnet was pivoted in the bottom of the glass so that it could swing around the rod.

On the right, a rigid wire was hung loosely from above, and a fixed magnet stuck up in the middle of the mercury. When the current was switched on the wire developed a magnetic field, which opposed the field from the magnet, so that on the left side the magnet twirled round the wire, and on the right the wire twirled around the magnet.

Simple electric motors

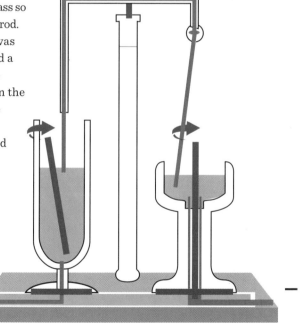

Faraday was excited by this, his first major discovery, and published his results without consulting or acknowledging either Davy or Wollaston. Wollaston was furious, claiming that Faraday had stolen his ideas, and a row ensued.

Davy died in 1829. Faraday was then free to carry on his research into electricity and magnetism, and he soon made what may have been his most important discovery, that a magnet could induce a current in a coil of wire: a phenomenon known as electromagnetic induction. He wrapped two coils of insulated wire around an iron ring; when he switched on a current in one wire it induced a momentary current in the other. Faraday also discovered that he could generate a current by moving a magnet through a loop of wire, and by moving a loop of wire over a stationary magnet. These experiments showed that a changing magnetic field produces an electric current; in other words, that mechanical energy could be converted into electrical energy. These discoveries would form the basis of the transformer and the dynamo.

Lines of force

Faraday had almost never been to school, and had no mathematical training. He was able, however, to describe magnetic fields in terms of lines of force.

He demonstrated these lines of force by placing a piece of paper over the poles of a magnet and sprinkling iron filings on the paper. The filings jump into a pattern of arcs, showing just how the field is arranged in space.

In 1845 Faraday showed that a strong magnetic field could rotate the plane of polarization of light, and he went on to discover the phenomenon of diamagnetism, in which some materials are weakly repelled by magnetic fields.

IS IT POSSIBLE TO STRETCH SOUND?

HOW MOVEMENT CHANGES THE PITCH OF A SOUND

THE STUDY

RESEARCHER:
Christian Andreas Doppler

SUBJECT AREA:
Acoustics

CONCLUSION:
Sound waves get compressed or expanded depending on our relationship to the source.

Born and raised in Salzburg, Austria, Christian Doppler was not strong enough to follow his father's career as stone mason. Instead he studied mathematics and physics, and in 1841 got a job at Prague Polytechnic in Bohemia.

Just a year later, at the age of 38, he published his most important work, 'On the coloured light of the binary stars and some other stars of the heavens', although the original was in German. In this paper he says that light is a wave motion and that its colour depends on the frequency of the wave.

Then he says that the frequency changes when the source or the observer moves, and he gives the analogy of a boat. A boat runs into waves much faster when going into the wind than when running before the wind; so the movement of the boat affects the frequency of meeting the waves. The same will happen, he says, with sound waves and light waves.

The Doppler effect

When an emergency vehicle – an ambulance, police car or fire engine – comes towards you, you hear the siren getting steadily louder, but notice the pitch, or note. As it passes you, the pitch of the siren slides down, and as the vehicle goes away the siren carries on at a lower note.

The reason this happens is that as the vehicle approaches you, the sound waves are bunched together. Each successive crest of sound is sent out a little closer to

The Doppler effect for a moving sound source

you than the previous one; so the crests are slightly closer together than they would be if the vehicle were stationary. If the crests are closer together, then the frequency has increased. When the vehicle is going away then each crest is sent out a little further away than the previous one, and the sound waves are stretched, meaning that the frequency is lowered.

By analogy, when a duck or a swan swims through water, the ripples in front of it are pushed together, and those beside and behind it are spread out.

Binary stars

In his 1842 paper, Doppler suggests that the natural colour of stars is white or pale yellow and, he argues, stars that are moving towards us will look more blue than usual, while stars that are moving away will look more red.

When two stars are close together they are called binary stars, and often they are spinning rapidly round each other. Albireo is a well-known binary star; the larger of the two is reddish; the smaller is distinctly blue. Doppler concluded that the larger one is moving away from us and the smaller one towards us.

In general he says that if the two stars are of equal brightness then the colours are complementary, but if they are of unequal brightness, then the brighter star is heavier, and the other is revolving around it. In the case of Albireo the larger star is only slightly red, while the other is very blue, implying that the blue star is doing most of the movement about a nearly stationary reddish partner.

He raises the example of periodic variable stars, which are invisible for most of the time but suddenly appear, looking red. He says they emit infrared radiation most of the time – so they are not visible – but they are binary stars, in orbit round an unseen partner, and at one stage in their orbit they speed up enough to shift the radiation into the red end of the spectrum, so that they become visible.

Astronomers today use the Doppler effect to measure the rate at which stars and galaxies are moving relative to us. Those moving towards us look more blue – they are blueshifted; those moving away are redshifted. In 1929 Edwin Hubble (see page 136) used the Doppler effect – the redshift of galaxies – to show that the universe is expanding.

In 1848 Hippolyte Fizeau discovered that the Doppler effect works also with electromagnetic waves; in France it is sometimes called the 'Doppler-Fizeau effect'.

Practical uses of the Doppler effect

Police officers use radar guns to catch motorists speeding. The radar waves are sent out of the radar gun, and bounce back off the car. The change of frequency tells the gun – and the operator – how fast the car is moving.

Doctors use a similar technique to measure blood flow, for example in the arteries of the neck, using ultrasound. Just holding the instrument against the neck at the correct angle allows them to measure the speed of the blood flow.

Vibration can be measured with a laser Doppler vibrometer. Point the laser at the surface in question, and from the reflected beam the instrument can work out the characteristics of the vibration.

1843

THE STUDY

RESEARCHER:
James Prescott Joule

SUBJECT AREA:
Thermodynamics

CONCLUSION:
You need a lot of energy to produce a little heat.

HOW MUCH ENERGY IS USED TO HEAT WATER?

THE NATURE OF HEAT

Way back in 1798 the curious American spy Count von Rumford, then working in Bavaria, had shown that a large amount of heat was generated when he tried boring cannons with a blunt drill. He thought that the heat was entirely caused by movement, and that it must correspond to some sort of particle movement in the iron.

Unfortunately most people thought heat was a fluid; when you put a hot thing next to a cold one, some of the fluid seeped into the cold thing and warmed it up. The French scientist Lavoisier called this fluid caloric (or rather *calorique*), and said that it could not be created or destroyed.

Steam or electricity?

Born in Salford in the north of England, James Joule followed his father into the brewing trade, but was enthusiastic about electricity, and did a variety of electrical experiments in his house. Wondering whether the steam engines in the brewery should be replaced by newfangled electric motors, he discovered in 1841 that 'the heat which is evolved by the proper action of any voltaic current is proportional to the square of the intensity of that current, multiplied by the resistance to conduction which it experiences'. We would write that as an equation: Heat is proportional to (current)^2 × resistance . This is called 'Joule's first law'.

Joule studied steam engines, and calculated that the amount of energy produced by even the best Cornish engines was equivalent to less than a tenth of the heat

generated in the boilers; so they were less than ten per
cent efficient – less efficient than a horse, he reckoned.

He noticed that in some of his electrical experiments,
parts of the circuit got hot. According to the caloric
theory the caloric must be coming from another part of
the circuit, since caloric could neither be created nor
destroyed – but Joule made careful measurements, and
nothing was cooling down. The electricity was definitely
making heat. What is more, if you pull a rope rapidly
through your closed fist, you can be seriously burnt. There
is no fluid involved – only movement. Joule decided to
investigate how much heat he could make by various
types of movement.

Paddle wheels

He made paddle wheels that just fitted inside
tanks of water, and then spun the wheels by
dropping weights on strings wrapped round the
spindles. He knew how much work had been done
by the falling weight, and he measured the
minute rise in temperature. It was so
small he had to do this repeatedly to get a
useful result.

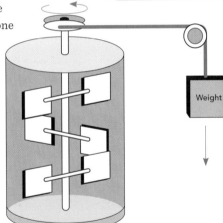

**Joule's
paddle wheel
experiment**

Weight

In one set of experiments he let the weight
fall 11 metres (36 feet), wound it up and
dropped it again 144 times – and the water
temperature rose just a few degrees.

Joule also heated water with electricity,
and by forcing it through narrow tubes.
According to legend, he spent his honeymoon
trying to measure the temperature difference
between the top and bottom of the Cascade de
Sallanches waterfall in the south of France. Unfortunately
the effect would have been very small; even Niagara Falls
only heats the water by about one fifth of a degree Celsius.

In all he heated water in five different ways, and

came to the conclusion that on average, to raise the temperature of 0.11 kilograms (0.25 pounds) of water by -17°C (1°F) you need to drop a 362 kilogram (800 pound) weight 30 centimetres (one foot).

The Joule-Thomson effect

Joule announced his results at the 1843 meeting of the British Association for the Advancement of Science, and was met by stony silence. His theories were controversial; as a result he had great trouble getting his work published in mainstream journals.

William Thomson, who later became Lord Kelvin, was sceptical, but they met again during Joule's honeymoon, and Thomson began to come around to Joule's point of view. Between 1852 and 1856 they communicated frequently, and together discovered the Joule-Thomson effect; gases forced under pressure through a valve cool down. This process is now the basis for all fridges, air conditioners and heat pumps.

Eventually, Joule's work gained widespread acceptance, and the SI unit of energy, the joule, is named after him.

Today we understand the mechanical equivalent of heat to be 4.2 joules per calorie.

DOES LIGHT TRAVEL FASTER IN WATER?

REFLECTION AND REFRACTION

1850
THE STUDY
RESEARCHERS:
Armand Hippolyte
Louis Fizeau and
Jean Bernard Léon
Foucault
SUBJECT AREA:
Optics
CONCLUSION:
Light definitely
travels in waves.

Ole Rømer had measured the speed of light in 1676 (see page 43), and James Bradley had adopted another astronomical method to measure it again in 1729, using what he called the 'aberration of light'.

Born five days apart in Paris, France, in September 1819, Hippolyte Fizeau and his friend Léon Foucault both became medical students, and they went together to a photographic course given by photography pioneer Louis-Jacques Daguerre. Together they worked on the improvement of photographic processes, but were overtaken by other experimenters and methods.

Measurement of the speed of light on Earth

At medical school Fizeau developed migraines; so he turned to physics. In July 1849 he devised a clever way to measure the speed of light directly, working in his parents' home in Paris. He spun a wheel with 100 teeth. Then he shone a beam of light through a gap between the teeth and reflected it back again from a mirror 8 kilometres (5 miles) away; so the light had to travel 16 kilometres (10 miles). He rotated the toothed wheel more and more rapidly until it was just going fast enough for him to see the light. This must mean that the light was going out through one gap and coming back through the next.

Fizeau's 1849 experiment

Mirror

Light source

Glass plate

Rotating disk

The problem was that light travels 16 kilometres (10 miles) in about a twenty-thousandth of a second, or 50 microseconds; so the gaps had to be very close together and the wheel had to spin very fast. Nevertheless in 1849 he calculated a speed of 313,000 kilometres per second (194,700 miles per second), which is about five per cent too high.

Léon Foucault also had to abandon his medical studies, because like the young Charles Darwin, he found he could not stand the sight of blood. In 1850 he and Fizeau joined forces, and together they built an even more clever system to measure the speed of light. Again light would be sent out on a long path – 32 kilometres (20 miles) each way this time, but first it would bounce off a rapidly revolving mirror.

When light came back from its 64-kiloemtre trip, the mirror would have rotated though a small angle; so that the return beam would be reflected at a slight angle A away from the source. From this angle A, and the speed of rotation of the mirror, they worked out the speed of light to be 298,000 kilometres per second or 185,000 miles per second, within one per cent of today's accepted value.

Fizeau and Foucault's 1850 experiment

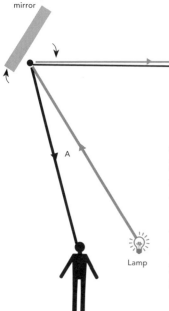

Rotating mirror

20 miles (32 km)

Mirror

A

Lamp

Observer

Speed of light in water

Foucault carried the experiment one stage further by inserting a tube of water in the path of the light, and showed that the light took longer to get back than it had in air.

Newton had predicted that light would travel faster through water than through air, because the dense medium would pull the particles of light through. In practice the speed of light in water turns out to be about 25 per cent lower than in air – 225,000 kilometres

(140,000 miles) per second. This result has been called 'the last nail in the coffin of the corpuscular theory' – and at last Thomas Young was proved right (see page 63).

The standard of length

In 1864 Fizeau suggested that 'the length of a light wave should be used as a length standard'. The speed of light in a vacuum (generally written as c) is now defined as exactly 299,792,458 metres per second (about 186,000 miles per second) and the metre is defined as the distance light travels in 1/2,997,924,58 seconds. In practice light travels about 0.3 meters (1 foot) in one nanosecond (one billionth of a second), while sound travels 0.3 metres (1 foot) in about one millisecond – a million times more slowly.

In water, however, light travels more slowly than in air, while sound travels much more quickly (see page 69).

Foucault's pendulum

On 3 February 1851, Foucault demonstrated the first ever proof that the Earth rotates. He suspended a heavy pendulum on a long chain in the Paris Observatory, after inviting all the scientists in Paris to come along. Later he hung it from the roof of the Panthéon in Paris. The pendulum is set swinging, and continues to swing in the same plane, relative to the stars. That means that as the Earth rotates, the plane of swing appears to turn; it can be used as a clock. This caused such public interest that Foucault pendulums were set up in major cities across the US and Europe.

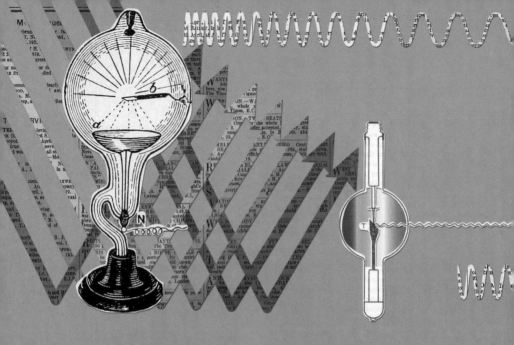

CHAPTER 4: Light, rays and atoms: 1851–1914

Physics and technology often go hand-in-hand. A new theory leads to a new piece of technology, which in turn enables new avenues of experiment and investigation. In the 17th century Torricelli's discovery of the vacuum led to the invention of the air pump, which in turn allowed Boyle and others to investigate the properties of the vacuum – or at least air at low pressure.

In 1865 Hermann Sprengel invented the mercury pump, which was more efficient than any previous machine, and allowed researchers such as William Crookes to send electrical discharges through what was almost empty space. This enabled the discovery of cathode rays, x-rays and the electron.

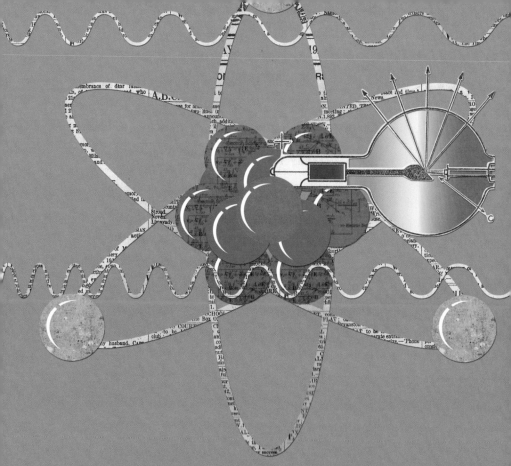

Then the discovery of x-rays sparked the discovery of radioactivity, and the remarkable work especially of Marie Curie enabled Ernest Rutherford to investigate the output of radioactivity, and label what he found as alpha, beta and gamma rays. Alpha rays turned out to be chunky particles – the nuclei of helium atoms – and he used them as missiles to investigate the structure of the atom. Beta rays and cathode rays turned out to be electrons, while gamma rays are the most energetic of all the waves in the electromagnetic spectrum.

1887

THE STUDY

RESEARCHERS:
Albert A Michelson
and Edward W Morley

SUBJECT AREA:
Cosmology

CONCLUSION:
There is no 'aether'.

WHAT IS THE AETHER?

THE RELATIVE MOTION OF THE EARTH AND LUMINIFEROUS AETHER

Sea waves travel in water; sound waves travel in air (or water); presumably light waves also need something to travel in. This is what scientists thought until the 1880s, and they called this something the 'luminiferous aether.' (Luminiferous means light-carrying.)

They knew that light could travel though a vacuum, as demonstrated by Torricelli (see page 34) and Boyle (see page 37), and also through space, since we can see the moon, sun and stars. Therefore it appeared that space, and vacuums on Earth, were permeated by this aether. But it was totally transparent, and appeared to show no frictional resistance to the movement of planets and moons; so does it really exist at all?

In its orbit around the sun, Earth moves at about 67,500 miles per hour (30 kilometres per second), and also spins on its axis; so whether the aether is stationary with respect to the universe, or with respect to the sun, or always moving through space, it must still be moving rapidly, relative to any particular spot on the surface of

Earth. So Michelson and Morley set out to investigate 'the relative motion of Earth and the luminiferous aether'.

Michelson tried his first experiments in Berlin in Germany in 1881, but the vibrations from traffic, even at 2am, made measurements difficult, and the apparatus was not sufficiently sensitive. He did, however, show that the method was manageable, and invented an interferometer, which he perfected after teaming up with Morley for the 1887 experiment at what is now Case Western Reserve University in Cleveland, Ohio.

The interferometer

The light from an oil lamp was focused on a half-silvered mirror, so that half of it went straight through, and half was bounced off at 90 degrees to the left. Each beam then bounced to and fro between a series of mirrors so that the path length, by the time the beams returned to the half-silvered mirror, was effectively 11 meters (36 foot), and when they reached the half-silvered mirror again, half of each beam arrived together at the telescope, where they formed interference fringes (see page 64).

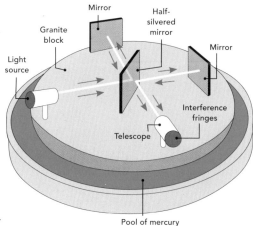

There was still a slight problem of vibration from occasional horse traffic and thunderstorms; so the whole of this apparatus was mounted on a massive three-ton block of stone. This in turn was floated in a pool of mercury, so that with a gentle push Michelson and Morley could turn the whole thing slowly through 360 degrees. Whichever way the aether (as it was called) was moving, at some point in the revolution of the apparatus, one of the beams would have to be going parallel to the motion of the aether, and the other at right angles. So they expected

Michelson and Morley's interferometer

to see a time difference in the arrival of the light beams, which would show up as a sideways movement of the interference fringes.

This was their idea: they had set up two beams of light, A and B, travelling at right angles to one another. When beam A was travelling across the flow of the aether, it should take less time to make the double journey than beam B, which was going parallel to the flow of the aether.

This is the same as a swimmer in a river; it takes less time to swim across the river and back than it takes to swim the same distance downstream and back up again; indeed if the stream is faster than the swimmer then the swimmer can never get back upstream.

At noon on 8 July, 1887, the researchers moved the apparatus steadily around in six complete circles, observing the interference fringes every sixteenth of a turn (22.5 degrees). They repeated this at 6 pm the same day; then they took noon and evening observations on two subsequent days.

They expected that at four points during each circle the fringes would shift sideways, first to the left, then to the right; so they expected to get a left-and-right pattern of movement, which they calculated should be at least twenty times as large as the smallest change they could measure.

The world's most famous 'failed' experiment

In fact they observed no movements at all. Michelson wrote in a letter to Lord Rayleigh: 'The Experiments on the relative motion of the earth and aether have been completed and the result decidedly negative.'

Did this mean that the aether was not moving at all at the surface of Earth? Perhaps it was being dragged along by all the clutter of trees and buildings. The researchers suggested that 'it is not impossible that at even moderate distances above the level of the sea, at the top of an isolated mountain peak, for instance, the relative motion might be perceptible.'

HOW WERE X-RAYS DISCOVERED?

SEEING THE SKELETON

1895
THE STUDY

RESEARCHERS:
Wilhelm Conrad
Röntgen, Antoine
Henri Becquerel

SUBJECT AREA:
Electromagnetic
spectrum and
radioactivity

CONCLUSION:
There is a
great variety of
electromagnetic
radiation, and some
heavy atoms are
unstable.

Excitement fizzed and crackled through the air of German and British scientific labs of the 1890s – and even more so through strange tubes, containing very little air. Vacuum pumps had been invented in the 17th century (see page 37), but by the 19th century were becoming much more powerful, and eventually could reduce the amount of air in a glass tube to about one millionth of normal atmospheric pressure.

Michael Faraday had noticed a strange arc of light between two electrodes (cathode and anode) in an evacuated glass tube in 1838, and in 1857 Heinrich Geissler, using a better pump, produced a glow that filled the tube, rather like a modern neon sign. In 1876 Eugen Goldstein showed that the rays would cast a shadow of a solid object in the tube; he called them 'cathode rays'. Then William Crookes, with a still more powerful pump, produced a glow, but noticed dark space in front of the cathode; it came to be called the 'Crookes dark space', and as he pumped more air out it spread down the tube to the anode, and then the glass behind the anode began to glow. He put this glow down to the cathode rays zooming down the tube and missing the anode to crash into the glass.

On Friday, 8 November 1895, Wilhelm Röntgen, then professor of physics at the

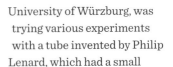

University of Würzburg, was trying various experiments with a tube invented by Philip Lenard, which had a small aluminium window sealed into the glass, in order to release some of the cathode rays. He chose for some reason to hold near this window a piece of cardboard painted with a fluorescent material (barium platinocyanide), and noted that it glowed brightly, even though there was apparently no light shining on it.

He then tried using a different tube, in complete darkness, and noticed a glow coming from across the room. He lit a match, and discovered it was the same fluorescent screen, which he had been planning to try next.

He was so excited he spent the entire weekend in the lab, repeating the experiments, and making sure he was not imagining this fluorescence. He had no idea what was causing it, but there had to be some sort of rays coming from the tube or the aluminium window, and he called them x-rays (x for unknown), although they came to be called Röntgen rays for some years.

After two weeks he took the first x-ray photograph – of the hand of his wife Anna Bertha. When she saw it she exclaimed 'I have seen my death.' Röntgen published his results 'On a new kind of rays' (*Über eine neue Art von Strahlen*) at the end of the year, and in 1901 he was awarded the first ever Nobel Prize in physics. He would not patent his discoveries, because he said he wanted everyone to benefit from them.

Within a month Röntgen's paper stimulated French physicist Henri Becquerel to investigate the phosphorescent salt potassium uranyl sulfate. Shine a light on a fluorescent material and it glows while the light shines on it; shine a light on a phosphorescent material and it goes on glowing after the light has been switched off. He thought this phosphorescent material might emit x-rays, or something like them.

He wrapped a photographic plate in two sheets of very thick black paper:

> such that the plate does not become clouded upon being exposed to the sun for a day. One places on the sheet of paper, on the outside, a slab of the phosphorescent substance, and one exposes the whole to the sun for several hours. When one then develops the photographic plate, one recognizes that the silhouette of the phosphorescent substance appears in black on the negative. If one places between the phosphorescent substance and the paper a piece of money or a metal screen pierced with a cut-out design, one sees the image of these objects appear on the negative. . . . One must conclude from these experiments that the phosphorescent substance in question emits rays which pass through the opaque paper and reduce silver salts.

Discovery of radioactivity

But then Becquerel discovered that he got the same results even without exposing the material to the sun. 'One hypothesis which presents itself to the mind naturally enough would be to suppose that these rays, whose effects have a great similarity to the effects produced by the rays studied by M Lenard and M Röntgen, are invisible rays emitted by phosphorescence.'

By May 1896 he realized that the cause of the new rays was the uranium in the phosphorescent material. By a superb stroke of serendipity Becquerel had discovered radioactivity.

THE STUDY

RESEARCHER:
Joseph John
Thomson

SUBJECT AREA:
Atomic physics

CONCLUSION:
The first clue about
what atoms are
made of.

WHAT'S INSIDE THE ATOM?

FINDING THE ELECTRON

In the breathtaking 1890s one experiment and one discovery led rapidly to others. Scientists eagerly watched their colleagues and pooled ideas for research. Electric light was only just becoming generally available, and motor cars were well into the future, but atomic science was burgeoning.

In the Cavendish Lab, in Cambridge, England, JJ Thomson, a physicist from Manchester, knew in 1897 that atoms were probably built from smaller particles, but the smallest of these was thought to be the same size as an atom of hydrogen, which is the lightest element (and the most abundant in the universe). Arthur Schuster had suggested in 1890 that cathode rays (see page 83) were negatively charged, and could be deflected by magnetic and electric fields. He estimated the ratio of charge to mass as more than a thousand, but no one really believed him.

Cathode rays

Thomson was also investigating cathode rays using an evacuated tube, and he noticed that they travelled much further through air than he would have expected for a particle the size of a hydrogen atom. Anything as big as that would soon bump into the molecules of nitrogen and oxygen in the air, and be brought to a shuddering halt, but the cathode rays seemed to escape these collisions.

Cathode rays spread out in all directions from the cathode, but he managed to corral a number of them into a narrow beam, which he was able to study in detail. He reckoned that the beam must be made of particles, because when it collided with a thermocouple it generated heat. In order to make quantitative measurements, he designed his tube so that the beam streamed out from the cathode, passed right through the anode, and into a bell jar, where it made a bright spot in the centre of a screen marked with a grid.

Bending the beam

The beam normally travelled in a straight line, but, like Schuster, Thomson found he could bend it not only with a magnet, but

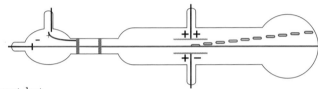

also with a powerful electric field, which showed that the beam must have a negative electrical charge. From the amount of bending he could calculate the ratio between the charge and the mass of the particles in the beam.

Deflection by electric field

The result was startling: the ratio of charge to mass for the cathode rays was more than a thousand times greater than it was for a hydrogen ion (H+), suggesting that each particle had a mass more than a thousand times less than a hydrogen atom (or was very highly charged). What was more, they seemed to have the same mass, regardless of which sort of cathode (i.e. atom) they came from. He concluded:

> As the cathode rays carry a charge of negative electricity, are deflected by an electrostatic force as if they were negatively electrified, and are acted on by a magnetic force in just the way in which this force would act on a negatively electrified body moving along the path of these rays, I can see no escape from

*the conclusion that they are charges of negative
electricity carried by particles of matter.*

Thomson called these particles 'corpuscles', although
they were soon called electrons, and he thought they must
be a part of every atom. In 1904 he proposed what came
to be called the 'plum-pudding model' of the atom: the
atom was a sphere of positive charge, with little electrons
embedded throughout, and probably orbiting rapidly.
His father had wanted Thomson to be an engineer, but the
family could not raise the cash for an apprenticeship;
so he studied science, went to Cambridge and
became a mathematical physicist. At the age of 28
he became Cavendish Professor of Experimental
Physics, which raised a few eyebrows, because not
only was he much younger than other applicants,
but also he had not done much experimental physics.
Thomson was, however, brilliant at designing
apparatus, and also a brilliant teacher. He won a Nobel
Prize in 1906, the second to come from the Cavendish Lab.
In all, this one laboratory has produced the remarkable
total of 29 Nobel Laureates.

Thomson and one of his students, FW Aston, also
investigated positive ions (atoms which had lost an
electron), and in 1912 found they could separate them
because of their various masses. One of their first
discoveries was that the rare gas neon has two isotopes
– that is atoms with the same number of protons but
different numbers of neutrons; we now call them neon-
20 and neon-22. The instrument Thomson and Aston
invented has evolved into the mass spectrometer,
one of the most powerful and useful machines at the
chemist's disposal.

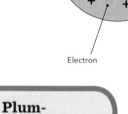

Spherical cloud of
positive charge

Electron

**Plum-
pudding
model**

HOW WAS RADIUM DISCOVERED?

PIONEERING THE STUDY OF RADIOACTIVITY

1898

THE STUDY

RESEARCHERS:
Marie Skłodowska-Curie and Pierre Curie

SUBJECT AREA:
Radioactivity

CONCLUSION:
The discovery of radium opens up the study of radioactivity.

Marie Curie was probably the greatest female scientist of all time. She had a tough childhood. Poland was a difficult place for nationalists in the late 19th century, and her family were hounded by the Russians, who also eliminated laboratory instruction from schools. Luckily her father was a physics teacher, and he brought much of the lab equipment home with him; so Maria Salomea Skłodowska, the youngest of his five children, was not totally deprived of an education.

She was able to enroll at the University of Paris, and met Pierre Curie, who was a lecturer in physics and chemistry, and managed to find her research space in his own lab.

Uranium rays

In late 1895 both x-rays and radioactivity were discovered (see page 83), and Marie (her French name) decided that she would investigate these mysterious 'uranium rays'. Luckily Pierre and his brother had developed an electrometer, which was a sensitive instrument for measuring electrical charge. Marie found that uranium rays caused the surrounding air to conduct electricity; so she could use the electrometer to detect the rays.

First she investigated various uranium salts, and found that the intensity of rays depended only on the quantity of uranium. This meant, she reckoned, that the rays were not some sort of molecule; they must be a property of the actual atoms of uranium.

One common ore of uranium is pitchblende (also called uraninite). Marie discovered that pitchblende produced four times as many rays as uranium metal, and deduced that pitchblende must contain some other material that was much more active than uranium. Looking for other active materials, she found in 1898 that thorium emitted rays also.

The new element

By this time Pierre had become fascinated by her research, and decided to join her, although she was definitely the driving force behind their joint investigations.

On 14 April 1898, they ground up and dissolved 100 grams (3.5 ounces) of pitchblende, hoping to find the new, highly active material. This was rather optimistic; eventually in 1902 they started with a ton of pitchblende, and after months and months of painstaking work they managed to isolate just 0.1gram (0.004 ounces) of radium chloride.

When they had extracted all the uranium from a sample of pitchblende by dissolving it in sulfuric acid, they found the residue was still radioactive. From it they managed to separate an element similar to bismuth – that is it lies below bismuth in the periodic table of the elements, and its compounds behave like bismuth's. This was a new element, never previously discovered, and Marie called it polonium, in honour of her native country. They announced the discovery in July 1898.

Unearthing the elusive radium

Then they carried on trying to sort out the rest of their material, and discovered something else highly active. It was similar to barium, and was all mixed up with barium compounds in the ore, but barium turns a flame bright green, and has green lines in its spectrum, whereas the new material gave unknown red lines; it had to be another new element. Separating it from barium was extremely difficult; all Marie and Pierre could do was make chloride salts and then crystallize them slowly; the chloride of the new material was slightly less soluble than barium chloride; so its crystals formed slightly sooner than barium's. They had to test every sample they collected with the electrometer, to see how radioactive it was. During this period they invented the term 'radioactivity'.

By 21 December 1898, they were confident that it was a new element. They called it radium, because of the abundance of rays it emitted, and on the 26th they announced its existence to the French Academy of Sciences, even though they had not yet isolated it in anything like a pure state. Marie finally isolated pure radium metal twelve years later. Radium compounds were crucial to the research that Ernest Rutherford did a few years later (see page 98); today the worldwide annual production of radium compounds is only about 100 grams (3.5 ounces).

Worldwide recognition

By 1902 Marie and Pierre had published a total of 32 scientific papers. In 1903 Marie was awarded her doctorate, and visited the Royal Institution in London, England, but women were not allowed to speak; so Pierre had to give the lecture, and afterwards when the audience asked questions he asked her; she replied to him and he repeated her answer to the audience.

In December that year Marie, Pierre and Henri Becquerel were awarded the Nobel Prize in Physics – she was the first woman so honoured. Initially the award was to be given only to Pierre and Becquerel, but when Pierre discovered this he complained, and they added her name too.

THE STUDY

RESEARCHER:
Nikola Tesla

SUBJECT AREA:
Electricity

CONCLUSION:
Electrical power
can be transmitted
without wires.

CAN POWER TRAVEL
THROUGH SPACE?

WIRELESS ENERGY TRANSMISSION

Nikola Tesla, born of Serbian parents in what is now
Croatia, was a mathematical prodigy at school. He ran
away from home to avoid the draft, went to the Austrian
Polytechnic and worked prodigiously hard, but then
became a gambling addict, failed his exams and ran away
again so that he did not have to admit his failure to his
family. Tall, good-looking and painfully thin, he became an
archetypal 'mad scientist'.

In June 1884 he went to New York to work for Thomas
Edison, but resigned from his job the following year after
a dispute over money that Tesla said Edison had promised
him. He managed to persuade various businessmen to
fund his research in exchange for a share in profits from
his patented inventions, and in 1888 signed a lucrative
contract with George Westinghouse.

In 1891 Tesla produced his most famous invention – the Tesla coil, which is a resonant transformer circuit that produces alternating current at enormously high voltages, and is still in occasional use today.

Wireless power transmission
At the 1893 World's Fair in Chicago, Westinghouse demonstrated the 'Tesla polyphase system', and one observer wrote:

> *Within the room was suspended two hard-rubber plates covered with tin foil. These were about fifteen feet apart, and served as terminals of the wires leading from the transformers. When the current was turned on, the lamps or tubes, which had no wires connected to them, but lay on a table between the suspended plates, or which might be held in the hand in almost any part of the room, were made luminous.*

In other words, the lamps demonstrated the transmission of electrical power without wires.

In 1899 Tesla set up a lab in Colorado Springs, because his polyphase alternating current system was installed there, and he had friends who gave him as much electricity as he wanted without the irritation of having to pay for it. In one of his first experiments he produced a 13-centimetre (5-inch) spark, for which he must have generated nearly half a million volts.

He started with a Tesla coil, and experimented with higher and higher voltages, up to four or five million

volts. He made artificial lightning with colossal sparks, and thunder that could be heard 24 kilometres (15 miles) away. People walking in the street found sparks jumping from their feet, horses bolted from a livery stable when they got shocks through their metal horseshoes and light bulbs glowed without being switched on. Tesla even managed to short out a power-station generator, which caused a major power outage.

His plan was to produce a 'magnifying transmitter', which he intended to use for wireless transmission of electrical energy, although he told people that he was working on transmission of radio signals. Tesla wrote 'I feel certain that of all my inventions, the Magnifying Transmitter will prove most important and valuable to future generations.'

The Wardenclyffe tower

In 1900, with backing from J Pierpont Morgan, Tesla began working on building a 57 metre (187 foot) high tower at Wardenclyffe, near Shoreham on Long Island, from where he hoped to transmit wireless signals and electrical power across the Atlantic. The tower was completed, but Tesla ran out of money, and the project died when Morgan refused to provide any more cash, having lost out badly in the 1901 stock market panic.

Tesla's best-known invention is the Tesla coil, but he took out dozens of patents, and invented all sorts of other electrical gadgets, including 'a plan to make dull students bright by saturating them unconsciously with electricity.'

As for wireless power transmission, it is now used on a small scale for charging electric toothbrushes, razors and cardiac pacemakers; for smartcards; and on a large scale for charging electric vehicles like buses and trains, including Maglev trains. Scientists and engineers are working on wireless chargers for mobile phones, smart tablets and portable computers. But Tesla's dreams have not been realized – yet.

IS THE SPEED OF LIGHT ALWAYS THE SAME?

E=MC²: THE THEORY OF SPECIAL RELATIVITY

What would you see if you could travel with a beam of light? Albert Einstein was born in Ulm, Germany, on 14 March, 1879. His parents moved to Italy in 1894, but Albert went to school at Aarau in Switzerland during 1895 and 1896, and found the teaching methods there were far more relaxed and progressive than at his previous school in Germany. He wrote much later 'It made an unforgettable impression on me, thanks to its liberal spirit and the simple earnestness of the teachers.' This was the time in which he says he began thinking about relativity.

Special relativity paradox
In his autobiography, Einstein describes a thought experiment (or *Gedankenexperiment*):

> . . .*a paradox upon which I had already hit at the age of sixteen: If I pursue a beam of light with the velocity c (velocity of light in a vacuum), I should observe such a beam of light as an electromagnetic field at rest though spatially oscillating. There seems to be no such thing, however, neither on the basis of experience nor according to Maxwell's equations. From the very beginning it appeared to me intuitively clear that, judged from the standpoint of such an observer, everything would have to happen according to the same laws as for*

1905

THE STUDY

RESEARCHER:
Albert Einstein

SUBJECT AREA:
Mechanics

CONCLUSION:
Special relativity is better than Newton's laws for describing mechanics at velocities approaching the speed of light.

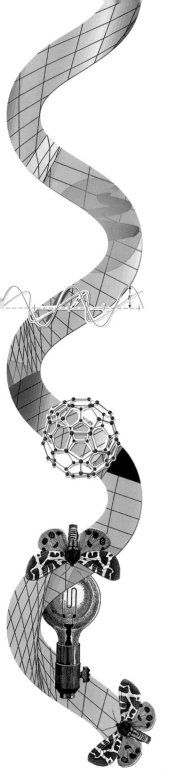

*an observer who, relative to the earth, was at rest.
For how should the first observer know or be able
to determine, that he is in a state of fast uniform
motion? One sees in this paradox the germ of the
special relativity theory.*

This is a paradox, because if Einstein saw the beam of
light at rest, he would know he was moving (at the speed of
light) – and this would have contravened Galileo's relativity.

Galileo explains, in his 1632 *Dialogue Concerning
the Two Chief World Systems*, that an observer in a
windowless cabin below decks on a boat in a calm sea has
no means of telling whether or not the boat is moving. He
can tell if the boat accelerates, or turns a corner, because
he will feel forces acting on him, but he cannot tell the
difference between motion in a straight line at a steady
speed, and staying still relative to the water.

Einstein may also have known about the Michelson-
Morley experiment (see page 80), which appeared to
show that the speed of light is not affected by the aether.
In any case, he started with the idea that the speed of
light is always exactly the same – 299,792,458 kilometres
per second (186,000 miles per second) or *c*.

This is counter-intuitive. Athletes generally take a run
up before throwing anything – cricket ball, javelin, football
– because running increases the speed of the object in
the air. Light is different; it is not affected by the motion
of the light source. It comes out of a torch at speed *c*,
regardless of whether the torch is stationary in your hand,
or travelling at high speed in a rocket.

In his 1905 paper on special relativity, Einstein also
assumed that the laws of physics are always the same in
any inertial frame of reference – that is in any vehicle or
place that is moving in a straight line at a constant speed.

There is no special place that is stationary, and
therefore no stationary aether for light waves to travel

in. Everything is moving, relative to everything else. You may think you are still, but relative to a Martian you are whirling through space.

Why does this matter?

The consequences of these ideas were profound. For one thing clocks could not be relied on to tell the same time in different frames of reference. If I can see you zooming past at high speed, then from my point of view your clock will be going more slowly than mine.

Also, events which appear to be simultaneous to one observer may not be simultaneous to another observer in a different frame of reference.

In that same year, known as his *annus mirabilis*, Einstein produced three more scientific papers: one on the photoelectric effect, which won him a Nobel Prize, one on the Brownian motion of molecules in a liquid and one on the equivalence of mass and energy, which was a direct development of special relativity, and is the basis of the world's most famous equation, $E = mc^2$.

In 1908 Hermann Minkowski, who had once taught Einstein, reformulated special relativity in terms not just of space, but including time. Einstein at first was sceptical of the idea of Minkowski's four-dimensional space-time, but later came not only to accept it, but to realize it was necessary for later work on general relativity.

The 1905 theory is called special relativity because it applies only to the case of observers in inertial frames of reference. When acceleration and gravity are involved, general relativity is needed (see page 116).

1908–1913

THE STUDY

RESEARCHERS:
Ernest Rutherford,
Johannes Wilhelm
Geiger and Ernest
Marsden

SUBJECT AREA:
Atomic physics

CONCLUSION:
Most of every atom is
empty space, with a
tiny dense nucleus in
the middle.

WHY IS THE WORLD MOSTLY EMPTY?

**THE ARTILLERY SHELL AND
THE TISSUE PAPER**

Ernest Rutherford, 'the father of nuclear physics', was
a farmer's son from New Zealand and a student of JJ
Thomson (see page 86). He won a Nobel Prize in 1908
for the work on radioactive decay that he did at McGill
University in Canada. This included discovering that
radioactive elements emit three sorts of 'rays': he
called them alpha, beta and gamma. When he moved to
Manchester, England, he showed that alpha 'rays' are
actually particles identical to the nuclei of helium atoms.
(We now know that these comprise two protons and two
neutrons, all bound together, with two positive charges.)

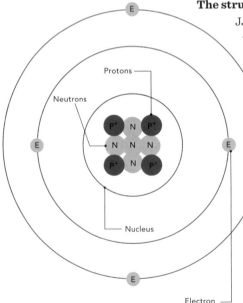

The structure of the atom

JJ Thomson had shown that electrons were
tiny particles of negative electricity, and
presumed that the rest of each atom
was a sphere of positive electricity,
with the electrons embedded in it –
the 'plum-pudding model'.

Rutherford decided to use alpha
particles as weapons to bombard
other atoms, in order to find out
something about their structure.
He invited a visiting German
scientist, Johannes Geiger, to do the
painstaking experiments, along with
Geiger's student Ernest Marsden.
In order to find out how many alpha

particles their radium source was producing, Rutherford and Geiger built a detector, in the shape of a glass tube containing air and a pair of electrodes. Each alpha particle would ionize some air, and create a pulse of electricity. This simple device evolved into the famous Geiger counter.

Rutherford was surprised by how much the alpha particles were scattered by the air they passed through, and suggested that Geiger and Marsden should investigate scattering by other materials. They decided to use gold leaf, because it was made of just one element and could be made extremely thin.

First they made a glass tube 2 metres (6 foot 6 inches) long, and at one end put a sample of radium, which emits alpha particles. In the middle of the tube was a slit 0.9 millimetres (0.04 inches) wide, allowing only a narrow beam to pass through. At the other end was a phosphorescent screen, which glowed when it was hit by an alpha particle; they used a microscope to count the scintillations (flashes) and to measure their spread. That meant spending hours in a dark room peering through the microscope and counting spots of light on the screen.

Scattered gold

When all the air had been pumped out of the tube, the scintillations formed a neat and narrow patch, but when air was let in the glow spread out. It was like shining a torch through a sheet of polythene. The same thing happened when there was no air but a piece of thin gold leaf over the slit; so both air molecules and gold atoms scattered the beam of alpha particles.

Rutherford calculated that if the atoms of gold were just diffuse spheres of positive charge, then the alpha particles should be deflected by very small angles; most of them would go straight through. So he was therefore quite surprised at the amount of the scattering, and suggested that next they should try to find out whether any of the particles were bounced off at large angles.

Big angles?

Geiger and Marsden built new apparatus, in which they shielded their screen with a slab of lead (which stops everything) and fixed their gold foil so that alpha particles could strike it at around 45 degrees and bounce off at the same sort of angle – just as you might use a mirror to peer round a partition. They did get some scattering, and found that gold scattered the particles more than aluminium, a much less dense metal.

From this and similar experiments, they deduced that the particles were deflected more by (a) thicker materials, (b) heavier atoms and (c) low-speed particles, but that a very small number were actually deflected by more than 90 degrees.

When they told Rutherford what they had found, he was astonished. In a lecture at Cambridge he said it was almost as incredible as if you fired a 38-centimentre (15-inch) shell at a piece of tissue paper and it came back and hit you.

On consideration, I realized that this scattering backward must be the result of a single collision, and when I made calculations I saw that it was impossible to get anything of that order of magnitude unless you took a system in which the greater part of the mass of the atom was concentrated in a minute nucleus. It was then that I had the idea of an atom with a minute massive centre, carrying a charge.

The point was that if the positive charge was all spread out, it would not have deflected the alpha particles much, but if it was a small solid lump then most particles would have missed it altogether, but a very few would have been batted back like cricketballs hitting a bat.

Rutherford concluded that the atom was mostly empty space, with a tiny positively charged nucleus in the centre, and presumably electrons whizzing round it.

HOW DO METALS
BEHAVE AT
ABSOLUTE ZERO?

**THE LINK BETWEEN SUPERCONDUCTIVITY
AND LOW TEMPERATURES**

1911
THE STUDY

RESEARCHER:
Heike Kamerlingh
Onnes

SUBJECT AREA:
Electricity

CONCLUSION:
Some metals become
superconductors
at very low
temperatures.

Weird things happen when the temperature approaches absolute zero. Robert Boyle (see page 37) discussed the possibility of minimum possible temperature, and later researchers saw that the volume of a fixed amount of gas steadily decreased as it was cooled. It looked as though the volume would shrink to nothing at around -270 °C (-455 °F).

After James Joule had sorted out the mechanical equivalent of heat (see page 72), Lord Kelvin calculated from thermodynamic principles that absolute zero would be -273.15 °C (-459.67 °F). Absolute temperature is now measured in either kelvins or rankines, in which absolute zero is 0 R (0 K) and the melting point of ice is 491.67 R (273.15 K).

Cryogenics

Dutch physicist Kamerlingh Onnes became professor of experimental physics at the University of Leiden, Netherlands, in 1882. In 1904 he built a large cryogenics lab in order to study low-temperature physics. There on 10 July 1908, he managed to liquefy helium gas at a temperature of 4.2 K; then, by pumping away the

remaining vapour, he reduced the temperature to 1.5 K. These were record-breaking temperatures at the time.

Lord Kelvin thought that the resistance of metals would increase enormously at such low temperatures; that electricity would simply stop flowing. Onnes, however, disagreed. On 11 April 1911, he dipped a solid mercury wire in liquid helium at 4.2 K, and its resistance disappeared completely. He was elated, and wrote in his notebook (which was not deciphered until 100 years later):

> *Mercury has passed into a new state, which on account of its extraordinary electrical properties may be called the superconductive state.*

This important breakthrough led to decades of low-temperature research, and many practical applications. For example the Large Hadron Collider (see page 169) uses 96 tons of helium to keep 1,600 superconducting magnets at 1.9 K.

Reaching absolute zero is not possible, but in 1999 a piece of rhodium metal was cooled to 0.000,000,000,1 K, which is close.

When liquid helium is cooled below 2.17 K it becomes superfluid, which means that if some is put in a cup or beaker, a thin film will climb up the walls and over the rim, until all the liquid has escaped. This is known as the Onnes Effect.

CAN YOU WIN A NOBEL PRIZE WITH YOUR HEAD IN THE CLOUDS?

CLOUD CHAMBERS AND THEIR IMPACT ON SCIENTIFIC DISCOVERY

1911
THE STUDY

RESEARCHER:
Charles Thomson
Rees Wilson

SUBJECT AREA:
Meteorology and
Particle physics

CONCLUSION:
The invention of the
cloud chamber leads
to unexpected
discoveries in physics.

Wilson's vision on a mountain top led to a breakthrough in particle physics. The son of a Scottish farmer, CTR Wilson planned to study medicine, but when he went to Cambridge University he became fascinated by physics, and in particular by meteorology.

In 1883, using money raised by public appeal, the Scottish Meteorological Society built an observatory on the top of Britain's highest mountain, the 1,344 metre (4,409 foot) Ben Nevis, near Fort William in Scotland. Every hour, resident meteorologists recorded rainfall, wind speed, temperature and so on, often in life-threatening conditions. Sadly the government refused to maintain the observatory, and it closed down in 1904.

Young physicists were sometimes employed there for a couple of weeks during the summer, to relieve regular staff. Wilson was delighted to be one of them, arriving in September 1894. Early one morning he stood near the highest point, with a sheer cliff dropping away in front of him. He was facing west, and as the sun rose behind him he saw his shadow on the clouds below. Then suddenly he saw the glory, or Brocken Spectre – a wonderful rainbow around the shadow of his head.

Amazed and delighted by this spectacle, he decided then and there to investigate the behaviour of clouds. Unfortunately he soon had to return to Cambridge, where the land is flat, and the clouds are less interesting; so he decided to build a cloud chamber – a flask in which he could make clouds artificially.

Clouds in a bottle
After a good deal of difficult glass-blowing Wilson managed to make a large glass flask with the right attachments.

He filled it with damp air, and then quickly reduced the pressure inside. The result was that the air became supersaturated with water vapour, and a few drops of water formed in the box, possibly on specks of dust. Disappointingly, Wilson found he could not make the sort of clouds he was interested in, but he wondered whether ionized air molecules might form cloud trails.

In late 1895 x-rays were discovered (see page 83), and early in 1896 Wilson tried beaming some into the cloud chamber. They caused an immediate mass of thick fog. Many years later he wrote 'I still vividly remember my delight at getting these results.' Clearly the x-rays had ionized some of the air – that is, knocked off electrons from some of the molecules, to leave positively charged ions – and the ions had acted as nuclei for the water drops to form.

He was able to do a little more research in the next few years, but between 1900 and 1910 he was busy teaching. But by 1910, he wrote, 'Ideas on the corpuscular nature of alpha- and beta-rays had become much more definite, and I had in view the possibility that the track of an ionizing particle might be made visible and photographed by condensing water on the ions which it liberated. . .'

Early in 1911 Wilson was able to get back to his cloud chamber, and found that charged particles did indeed leave tracks, like the vapour trails of aircraft. This was the first time anyone had been able to see them. He was soon able to photograph tracks of individual electrons and alpha particles. The electrons, he said, made 'little wisps and threads of cloud.'

The most amazing discovery
In 1923 Wilson finally perfected the operation of his cloud chamber and produced two beautifully illustrated papers on the tracks of electrons. This stimulated worldwide interest, and soon cloud chambers were in operation in Paris, Leningrad, Berlin and Tokyo. Cloud chambers demonstrated the discovery of the positron, the annihilation of electron and positron, and the transmutation of atomic nuclei. They also allowed physicists to study cosmic rays (see page 138). Rutherford (see page 98) said that the cloud chamber was 'the most original and wonderful instrument in scientific history.'

In 1927 CTR Wilson won the Nobel Prize in physics, 'for his method of making the paths of electrically charged particles visible by condensation of vapour' – even though he had invented the idea for an entirely different discipline. He himself wrote 'the whole of my scientific work undoubtedly developed from the experiments I was led to make by what I saw during my fortnight on Ben Nevis in September 1894.'

1913

THE STUDY

RESEARCHERS:
Robert Andrews
Millikan and Harvey
Fletcher

SUBJECT AREA:
Particle physics

CONCLUSION:
The charge is
1.6 × 10⁻¹⁹ coulomb.

CAN A PARTICLE'S CHARGE BE MEASURED?

TESTING THE ELECTRON

The electron had been discovered by JJ Thomson in 1897 (see page 86), and he had measured the ratio of the electron's charge to its mass, but no one knew the actual value of either; so if the charge could be measured then the mass could be calculated.

Robert Millikan became a professor at the University of Chicago in 1910, by which time he had started his experiments with oil drops. With the help of his graduate student Harvey Fletcher he devised and set up what was in essence a simple experiment.

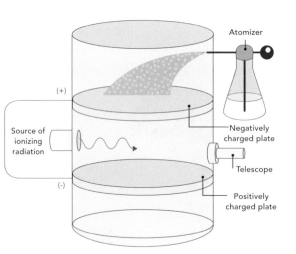

Atomizer

(+)

Source of
ionizing
radiation

Negatively
charged plate

Telescope

(-)

Positively
charged plate

Measuring the very small

Using a scent atomizer, they blew minute droplets of oil into the reservoir above an observation chamber, and then watched through a telescope to see how fast they fell through the air.

Then they let in a burst of x-rays, which ionized some of the air in the chamber, by knocking off electrons to make the molecules positively charged. When one of these ionized molecules collided with an oil droplet, the positive charge was

**Measuring
an electron**

transferred to the oil. This made no difference to the effect of gravity, but then the researchers applied an electric field.

There were metal plates above and below the chamber, which could be charged up to a maximum of 5,300 V, positive below and negative above. This electric field acted against gravity, pushing the oil drop upwards away from the positive plate and towards the negative plate. The researchers were able to see whether the drop was still falling, stationary or moving upward and to measure how fast.

They did not know how many charges had been transferred to each drop, but they assumed that there was a fundamental unit of charge, and that the total charge on any one drop would be a multiple of this – perhaps 2, 4 or 5 times the unit.

They knew the viscosity of air and the temperature of each trial, and how the effect of viscosity varied with

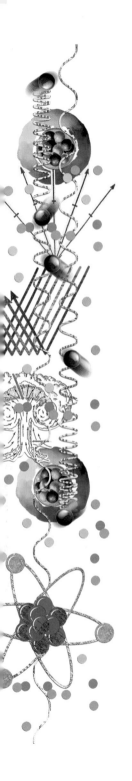

extremely small droplets. Therefore from the rate at which it fell, they could calculate the effective weight of each drop.

Electric fields

Then they switched on the electric field, and carefully varied it until the drop was neither rising nor falling. This was slow and difficult work. In all they studied 58 drops, and watched some of them for five hours. When the drop remained stationary, they knew that its weight was exactly matched by the upward force exerted by the electric field, which they could calculate from the voltage they had applied. Since they already knew the weight of the drop they could calculate the charge it was carrying.

They also applied a larger positive field, and observed the drops 'falling upwards.' From the rate at which the drops moved, the researchers could again measure the charge they were carrying.

Putting together the results from many drops, they reached the conclusion that the fundamental unit of charge must be 1.592×10^{-19} C, while the accepted value today is 1.602×10^{-19} C; so Millikan and Fletcher were within one per cent of today's value; their small error was probably the result of their incorrect value for the viscosity of air.

Discoveries

This was an important result for several reasons. First, it established that electrical charge comes in discrete units and is not a continuous variable as Thomas Edison and many others believed.

Second, if this was the smallest possible charge, it must be the charge on a single electron.

Third, it provided a value for Avogadro's number, named after the Italian scientist Lorenzo Romano Amedeo Carlo Avogadro di Quaregna e di Cerreto, Count of Quaregna and Cerreto, who suggested in 1811 that the volume of any sample of gas (at a given temperature and pressure) is proportional to the number of particles (atoms or molecules). Avogadro's number is the number of atoms in 1 gram (0.035 ounces) of hydrogen, 12 grams (0.42 ounces) of carbon, 16 grams (0.56 ounces) of oxygen, or 56 grams (1.98 ounces) of iron; it is 6×10^{23}, or 600 thousand million million million.

Millikan aroused some controversy by leaving out half his results. This sort of data massaging is bad practice, and can lead to downright fraud. In fact this did not make any difference to Millikan's result, but had he included all the data the statistical error would have been larger.

Naturally Harvey Fletcher, the graduate student, did the vast majority of the actual tedious observation of floating drops through a telescope. However, in a most unusual agreement he allowed Millikan to take sole credit for the work in the published papers, in exchange for being able to take sole credit for a related result in his PhD dissertation. As a result Fletcher got his PhD, and Millikan won the Nobel Prize in physics in 1923.

Millikan did not believe Einstein's 1905 paper on the photoelectric effect and carried out a long series of difficult experiments trying to prove him wrong, only to prove him right. He said, 'I spent ten years of my life testing that 1905 equation of Einstein's, and contrary to all my expectations, I was compelled in 1915 to assert its unambiguous verification in spite of its unreasonableness.'

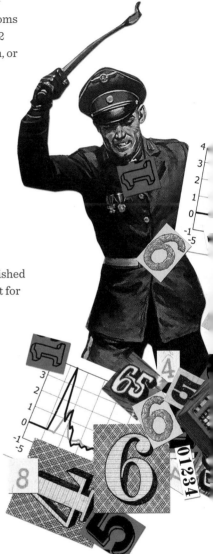

THE STUDY

RESEARCHERS:
James Franck and
Gustav Ludwig Hertz

SUBJECT AREA:
Quantum mechanics

CONCLUSION:
The quantum
theory of mechanics
was practically
demonstrated for
the first time.

IS QUANTUM MECHANICS STRANGER THAN WE IMAGINE?

THE QUANTUM LEAP

What effect would floating mercury atoms have on flying electrons? Franck and Hertz worked together at the University of Berlin. In their first paper, presented on 14 April 1914, they describe how they generated electrons from a cathode, and passed them down an evacuated tube through a metal mesh grid to the anode.

Electrons are negatively charged; so they were attracted to the positively charged grid, and travelled faster if the positive voltage on the grid was increased. The anode carried a small negative charge, relative to the grid; so the electrons could reach it only if they were travelling fast enough.

The tube contained mercury vapor; there was a drop of liquid mercury inside, and the tube was heated to 115 °C (239 °F). The electrons flying through the tube were therefore liable to collide with mercury atoms floating about in their path.

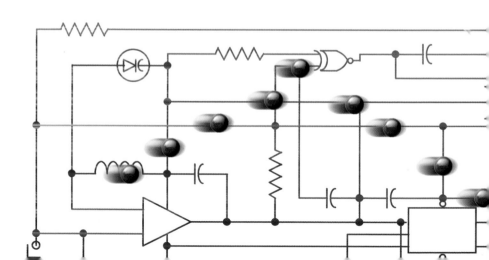

The researchers measured the current – the flow of electrons reaching the anode. They found that as they increased the voltage on the grid, the current gradually increased until the voltage reached 4.9 V, when the current suddenly dropped, almost to zero. This meant that the speed of the electrons had increased steadily to 1.3 million meters per second (4.3 million feet per second), and then suddenly dropped to nothing.

When the researchers increased the voltage on the grid, the current climbed again until the voltage reached 9.8 (i.e. 2 x 4.9), when the current suddenly dropped again. The same thing happened at 14.7 V (3 x 4.9 V).

Apparently the electrons could lose only 4.9 electron volts of energy, no more, no less. An electron going faster than the critical speed lost only 4.9 eV, and then carried on. Franck and Hertz pointed out that this 4.9 eV corresponded to one of the many lines in the spectrum of mercury atoms, at 254 nanometres (nm).

What was going on?

At first Franck and Hertz thought that the mercury atoms were being ionized by the flying electrons, but then Niels Bohr stepped in, with a new model of the atom. He had published this the previous year, but Franck and Hertz had not seen it.

The 'plum-pudding' model of JJ Thomson (see page 88) had been superseded by Rutherford's model (see page 98) of

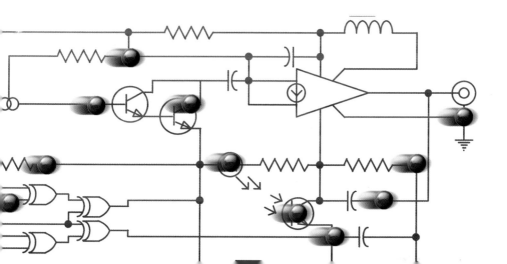

a small nucleus surrounded by empty space, with electrons whizzing about, perhaps orbiting the nucleus. There was a big problem with this, however. Whizzing electrons should emit light, and atoms don't emit light. Also, the negatively charged electrons should crash into the positively charged nucleus, but they don't.

Continuous streams of energy?

German physicist Max Planck had suggested that energy might come not in a continuous stream, but in discrete packets, or 'quanta'. In his 1905 paper on the photoelectric effect, Einstein showed that this was true of light.

In Copenhagen, Niels Bohr wondered whether a similar idea might apply to electrons. He proposed a new model of the atom in which the electrons were whizzing about the nucleus, but in fixed energy levels (he called them 'stationary orbits'). There would be a maximum of two electrons in the lowest energy level, and they could not get closer to the nucleus than their fixed stationary orbit. In the next energy level there would be a maximum of six electrons, and so on. All the levels were quantized – fixed in size and energy, rather like the quanta of light. Electrons could be bumped up to a higher energy level (if there was a vacant space) by a specific amount of energy, and if they then dropped back they would emit the same amount of energy.

Bohr pointed out that the 4.9 V change observed by Franck and Hertz corresponded to a difference between two quantum energy levels of the mercury atom; so what was probably happening was that electrons within the mercury atoms were being excited to a higher energy level. He also suggested that when these electrons dropped

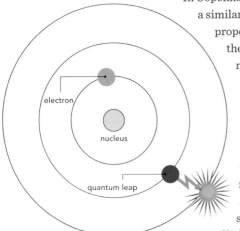

electron

nucleus

quantum leap

The Bohr effect

back to their original level they should emit ultraviolet light at a wavelength of 254 nm.

In their second paper, in May 1914, Franck and Hertz reported that under the conditions of their experiment, the mercury emitted light almost entirely at 254 nm, therefore confirming that the excited atoms were returning to their 'ground state'.

Making sense of the results

Now the original results made sense. The electrons in the mercury atoms could not be excited at any voltage less than 4.9, because 4.9 eV was the smallest gap between the quantized energy levels that were full and the next empty level. When the voltage was lower than 4.9, the electrons simply bounced off the mercury atoms, and carried on towards the grid and the anode. When the voltage reached 4.9, however, most of the electrons walloped into mercury atoms with enough energy to excite the mercury, leaving the incoming electron without enough energy to finish the journey to the anode; so the current dropped almost to zero.

When the voltage reached 9.8, almost every electron collided with two mercury atoms, one after the other, and excited both of them, before giving up the journey. Once again the voltage dropped almost to zero.

All the excited mercury atoms soon started glowing at 254 nm, as the excited electrons dropped back to their original quantum level.

This therefore was the first experimental evidence in support of the fledgling theory of quantum mechanics. After a presentation of these results by Franck a few years later, Albert Einstein is said to have remarked, 'It's so lovely it makes you cry.' It proved that electrons can appear and disappear in different orbits without 'travelling' to their new destinations. This is the famous 'quantum leap'.

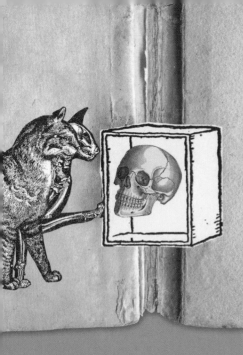

CHAPTER 5: Deeper into matter : 1915–1939

At the close of the 19th century one of the Grand Old Men of Physics, Lord Kelvin, is reputed to have said 'There is nothing new to be discovered in physics now.' Within a few years special relativity and quantum mechanics changed the world.

In the 20th century, physics became stranger and weirder. In 1915 Einstein showed how gravity can warp space-time; Rutherford fulfilled the alchemist's dream, by turning one element into another; and a Belgian priest suggested that the universe had a beginning in the shape of a cosmic egg.

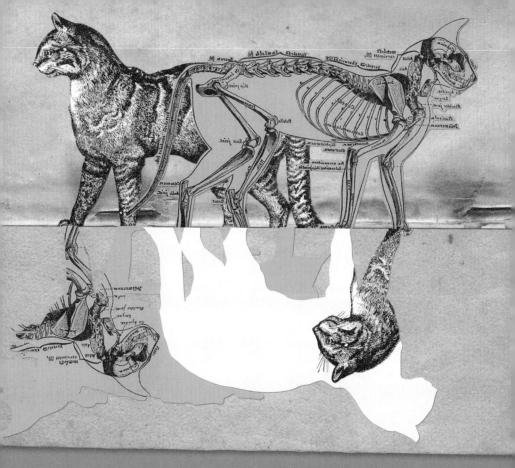

Aristocratic French physicist Louis de Broglie outrageously suggested that electrons might also behave as waves, and at Bell Labs, Davisson and Germer proved that this was true: electrons are both particles and waves. Then, Paul Dirac predicted the existence of antimatter, and it was discovered in 1932 by Carl Anderson at Caltech.

   Instead of 'more and more precise measurement', Heisenberg showed that at the atomic scale it was and always would be impossible to measure precisely both the position and the velocity of anything. Physics would forever be uncertain.

THE STUDY

RESEARCHER:
Albert Einstein

SUBJECT AREA:
General relativity

CONCLUSION:
Clocks and light are
affected by gravity.

IS GRAVITY RELATED TO ACCELERATION?

EINSTEIN'S THEORY OF GENERAL RELATIVITY

Galileo had shown (see page 31) that big things and small things fall at the same speed. Imagine that you dropped tomatoes in a lift, but just as you dropped them, the elevator cable broke. The lift would fall too, along with you and the tomatoes, and you would all fall at the same speed; so the tomatoes would stay where they were, right by your hands. You are all in free fall.

An astronaut in a spacecraft in orbit around Earth is also in free fall. She may feel weightless, but in fact gravity is just pulling her and the spacecraft towards the Earth strongly enough to keep them in orbit.

When the rocket motors fire up, she will feel pushed down towards the back of the craft, just as she was pushed down by gravity before it took off from Earth. In fact the effects of gravity are exactly the same as the effects of acceleration. This is Einstein's 'equivalence principle'; he called it his 'happiest thought'.

At the back of the spacecraft is a curious clock – a strobe light, flashing ten times every second. When the craft is stationary, lying down horizontal on Earth, the

At rest on
Earth's
surface

In
rocket

flashes reach the front of the craft ten times every second. When the craft is accelerating through space, however, their arrival frequency decreases. They are still leaving the back at ten times per second, but between each pair of flashes the craft gains a bit more speed, and so the flashes take longer and longer to reach the front; they may arrive at a rate of only nine per second.

So, in this accelerating frame of reference (the spacecraft), from the point of view of an observer at the front, the clock at the back is running slow. The signal is subject to gravitational redshift (see page 136).

Because acceleration and gravity have the same effects, a clock runs slow in a strong gravitational field: this is 'gravitational time dilation'.

Conversely, if the strobe were flashing from the front of the craft under either acceleration or gravity, an observer at the back would see the clock running fast – a gravitational blueshift. Einstein published his theory of gravitational frequency shift in 1915, and it has been verified by various experiments. In 1960 Robert Pound and Glen Rebka passed gamma rays up and down a 22 metre (72 foot) tower, and showed that their frequency was shifted as predicted.

Using atomic clocks

A more dramatic test was carried out in October 1971 by physicist Joseph Hafele and astronomer Richard Keating, who worked with atomic clocks at the US Naval Observatory. They took four ultra-precise atomic clocks around the world in commercial airliners, first eastwards and then westwards. Then they compared the time with that shown by atomic clocks at the Naval Observatory.

Compared with a frame of reference at rest with respect to the centre of Earth, general relativity predicted that all the clocks in the air should have gone faster than those on the ground, because gravity is lower at 9,000–12,000 metres (30–40,000 feet).

Meanwhile, special relativity (see page 95) predicted that clocks going eastwards, in the same direction as the Earth's surface, would be moving faster than those on the ground, and would therefore run more slowly. Those moving westwards would be moving more slowly than the clocks on the surface, and would therefore run faster. The net result of these various effects were that the eastward clocks were predicted to lose around 50 nanoseconds (ns, billionths of a second), relative to those at the USNO, while the westward clocks were predicted to gain about 275 ns, and that is just what happened.

Gravity and light

Imagine our astronaut in her spacecraft in orbit. They are in free fall, and so the spacecraft is an inertial frame (see page 96). She fires an arrow across the spacecraft and hits the target on the opposite wall. But if the rockets fire and the craft accelerates as she fires, then the ship will move forward, and the arrow will miss the target and hit the wall nearer the back of the craft. The same thing will happen if she fires a laser beam across the craft; in free fall it will go straight, but under acceleration the beam will appear to bend, and if the acceleration is high enough, it also will miss the target.

Because a gravitational field is equivalent to acceleration, gravity will also bend a light beam; if she fired her laser across the craft while it was sitting on the launch pad before lift-off, the beam would still bend downwards under the Earth's gravity, although by only a microscopic amount.

Einstein suggested a startling new idea: there is no force of gravity. Instead, near a massive object like Earth, space-time (see page 97) itself is curved, so that the natural motion of the spacecraft and the astronaut is not Newton's straight line, but an orbit.

CAN YOU TURN LEAD INTO GOLD?

LIMITATIONS OF TRANSMUTING ELEMENTS

1919
THE STUDY

RESEARCHER:
Ernest Rutherford

SUBJECT AREA:
Atomic physics

CONCLUSION:
Elements can be transmuted, but not lead into gold.

Fresh from his triumph in using alpha particles (aka helium nuclei) to find a new structure for the atom (see page 98), Ernest Rutherford used the same missiles to turn nitrogen into oxygen.

He had noticed that the alpha particles don't penetrate far through air, and he found that as they ran into air molecules, some curious radiation was emitted, which:

Gives rise to a number of scintillations on a zinc sulphide screen far beyond the range of the alpha particles. The swift atoms causing the scintillations carry a positive charge and are deflected by a magnetic field and have about the same range and energy as the swift hydrogen atoms produced by the passage of alpha particles through hydrogen . . .

The intense source of radium C was placed inside a metal box about 1.2 in [3 cm] from the end, and an opening in the end of the box was covered with a silver plate of stopping power equal to about 2.4 in [6 cm] of air. The zinc sulphide screen was mounted outside about 0.04 in [1 mm] distant from the silver plate to admit of the introduction of absorbing foils between them.. . . The air in the box was exhausted.. . . When dried oxygen or carbon dioxide was admitted into the vessel the number of scintillations diminished to about the same amount to be expected from the stopping power of the column of gas.

*A surprising effect was noticed, however, when dried
air was introduced. Instead of diminishing, the number
of scintillations was increased, and for an absorption
corresponding to about 7.5 in [19 cm] of air the number
was about twice that observed when the air was
exhausted. It was clear from this experiment that the
alpha particles in their passage through air gave rise to
long-range scintillations which appeared to the eye to be
about equal in brightness to hydrogen scintillations.*

Rutherford already knew that oxygen did not produce
these scintillations, and 99 per cent of air is a mixture
of oxygen and nitrogen; so it seemed probable that the
radiation came from collision of alpha particles with
nitrogen molecules.

Smashing nitrogen atoms
He therefore bombarded pure nitrogen gas with alpha
particles, and among the products he again observed
the nuclei of hydrogen atoms. We would call them H⁺ or
protons, but the proton had not yet been discovered or
named; so he called them hydrogen nuclei. They must
have been knocked out of the nuclei of nitrogen atoms by
the collisions. He found it

> *... difficult to avoid the conclusion that these long-range
> atoms arising from the collision of alpha particles with
> nitrogen are not nitrogen atoms but probably charged
> atoms of hydrogen.... If this be the case, we must conclude
> that the nitrogen atom is disintegrated under the intense
> forces developed in close collision with a swift α particle,
> and that the hydrogen atom which is liberated formed a
> constituent part of the nitrogen nucleus.*

In other words, that hydrogen nuclei were one of
the constituents of nitrogen nuclei, and possibly of

all atomic nuclei. This seemed at least possible, since hydrogen is the lightest element, and most elements have atomic masses that are approximately equal to multiples of the mass of hydrogen atoms.

> *Considering the enormous intensity of the forces brought into play, it is not so much a matter of surprise that the nitrogen atom should suffer disintegration as that the α particle itself escapes disruption into its constituents. The results as a whole suggest that, if α particles – or similar projectiles – of still greater energy were available for experiment, we might expect to break down the nuclear structure of many of the lighter atoms.*

After moving to Cambridge, Rutherford asked PMS Blackett to investigate the reaction of alpha particles and nitrogen, using a cloud chamber. By 1924 Blackett had taken 23,000 photographs, which showed 415,000 tracks of ionized particles. Eight of these tracks showed that the alpha-nitrogen collision produced an unstable fluorine atom, which then decayed into an oxygen atom and a proton. [N + He → [F] → O + H]

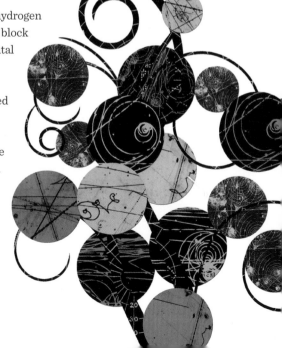

In 1920 Rutherford decided that a hydrogen nucleus could well be a basic building block of all nuclei, and also a new fundamental particle; he called it the proton.

The following year, while working with Niels Bohr, Rutherford suggested the possibility that there were also electrically neutral particles in most atomic nuclei, which would dilute the repulsive positively charged protons. He proposed that they might be called neutrons.

1919

THE STUDY

RESEARCHERS:
AS Eddington,
FW Dyson, and
C Davidson

SUBJECT AREA:
Astrophysics

CONCLUSION:
Einstein was right.

COULD EINSTEIN BE PROVED RIGHT?

PRACTICAL CONFIRMATION OF THE THEORY OF GENERAL RELATIVITY

Born in Kendal, England, in 1882, Arthur Eddington, a Quaker and pacifist, became a professor of astronomy at Cambridge at the age of 31. He was intuitive, and had a tendency to make guesses about such topics as the structure of stars and where their energy comes from – and then look for evidence to back his hunches. Often he was proved right.

When Eddington heard about Einstein's theory of general relativity (see page 116) he was excited. Britain and Germany were at war at the time; so no jingoistic Englishman would be interested, but he was a pacifist, and became the champion of relativity in Britain.

As a result he joined forces with the Astronomer Royal Frank Watson Dyson, and they persuaded the government to fund two expeditions to observe the total solar eclipse of 29 May 1919, in order to gather evidence which might support Einstein's theory.

The prediction

General relativity predicted that gravity would bend rays of light. If light from a distant star passes close by the sun, then the rays of light should be pulled in towards the sun by its huge gravitational field, and so the star should appear to be in slightly the wrong position.

Most of the time this is impossible to observe, because light from the sun prevents observation of distant stars that appear just past its edge (or 'limb'). During a total eclipse, however, the sun's light is blocked by the moon;

so for a few minutes those stars are visible. Photographs taken during totality can be compared with others taken later, and the positions of the distant stars checked.

The effect was expected to be small. If the theory was correct the light would be bent by a minute angle. The circle is divided into 360° (degrees); each degree is divided into 60′ (minutes); each minute is divided into 60″ (seconds). The stars should appear further out from the sun than they really are. Newtonian gravity predicted that the starlight would be bent by 0″.87 (i.e., less than one second); Einstein predicted that the light would be bent by 1″.75 – exactly twice as much.

Where in the world?

The path of the eclipse would run from Brazil across the Atlantic and across the middle of Africa to Lake Tanganyika. The researchers decided to send expeditions both to Sobral in Brazil and to the Portuguese island of Principe, in the Gulf of Guinea, west of central Africa.

They collected the best available telescopes, had special collapsible huts made and embarked on the *Anselm* on 8 March 1919. The Brazil team arrived on the Brazilian coast on 23 March and eventually proceeded to Sobral by steamer and train.

Meanwhile, the Principe team travelled on the *Anselm* as far as Madeira, then transferred to the *Portugal*, and reached Principe on 23 April. They set up their equipment in a small walled enclosure looking west, down towards the ocean.

The day of the eclipse

In Brazil the morning was cloudy, and at the time of 'first contact' (when the moon begins to cross the face of the sun) there was nine-tenths cloud, but they could just see enough of the sun to get their

telescopes lined up accurately. Then gradually the clouds separated, and one minute before totality they had a clear space around the sun. As the sun disappeared they started a metronome, and one of them called out every tenth beat; that was how they timed their exposures. Using two cameras they exposed a total of 27 photographic plates.

The Principe team were less fortunate. There was a heavy thunderstorm on the morning of the eclipse, which was due at 2:15 p.m. There was thick cloud most of the morning, and then by 1:55 they could see the sun through drifting cloud. They managed to expose 16 photographic plates, but only seven were useful.

They were nearly marooned for months on the island by a strike in the steamship company, but managed to return to England on 14 July.

Results and conclusion

The paper describing their expedition is 45 pages long, and contains many pages of tables and calculations, but the overall results for the bending of starlight were:

Brazil   1″.98 ± 0″.12
Principe 1″.61 ± 0″.30

These results are both much closer to the prediction from general relativity of 1″.75, rather than the Newtonian value of 0″.87, and therefore were strong evidence for Einstein's theory. (The ± numbers are the estimated errors; so they calculated the Brazil result to be between 1″.86 and 2″.10.)

During the 1920s and 1930s Eddington wrote many popular books on relativity, atoms, stars and cosmology, and gave dozens of lectures and interviews, as well as becoming a radio star and a household name.

DO PARTICLES SPIN?

THE STERN-GERLACH EXPERIMENT

1922
THE STUDY

RESEARCHERS:
Otto Stern and
Walther Gerlach

SUBJECT AREA:
Atomic physics and
Quantum mechanics

CONCLUSION:
Electrons spin in just
two ways.

Around 1920 there was some controversy among
scientists about the newfangled quantum mechanics and
the structure of the atom. In the classical (Rutherford)
model the negatively charged electrons were zooming
around the positively charged nucleus. This should mean
that they behave as tiny magnets; this had been known
since the time of Faraday (see page 66).

If a beam of atoms is passed through a magnetic field,
it should therefore be bent, because the tiny magnets
will be attracted or repelled by the field, provided that
the magnetic field is not homogeneous (all the same), so
that in effect the North Pole is stronger than the South
(or vice versa). The atoms might be in any orientation; so
they should be bent equally in all directions. Therefore
the classical theory would predict that the beam would be
spread out in all directions. If it hits a screen it will form a
broad patch.

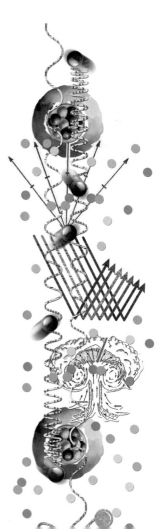

The value of particles' spin

Niels Bohr, pioneer of quantum mechanics, said that the
magnetic moment (or 'spin') of such a particle could have
only two values: $+\frac{1}{2}$ or $-\frac{1}{2}$. The orientation of the atom
should not make any difference. This is the quantum
nature of spin. If this were correct, then the beam should
be split into two, and produce two spots on the screen.
Otto Stern, a German Jew born in what is now Poland,
worked with Einstein and went to Frankfurt in Germany
in 1915. Walter Gerlach, also a German physicist, served
in the German army in World War I, and in 1921 became a

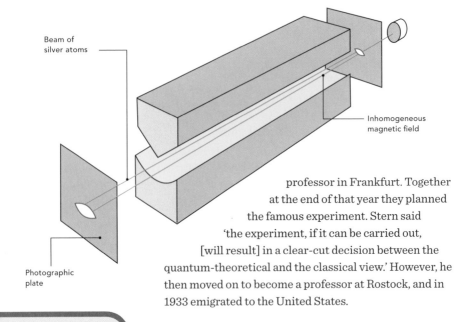

Beam of
silver atoms

Inhomogeneous
magnetic field

Photographic
plate

professor in Frankfurt. Together
at the end of that year they planned
the famous experiment. Stern said
'the experiment, if it can be carried out,
[will result] in a clear-cut decision between the
quantum-theoretical and the classical view.' However, he
then moved on to become a professor at Rostock, and in
1933 emigrated to the United States.

Testing the spin of nuclear atoms

A triumph

At the University of Frankfurt in early 1922 Gerlach
fired a beam of silver atoms through the magnetic field.
The latest theory, by Bohr and Sommerfeld, was that the
nuclei of silver atoms should have spin.

When the field was homogeneous the beam made
a fat stripe on the screen. When he made the field
inhomogeneous the fat stripe spread out in the centre to
make two lines, which came together top and bottom to
make what looked like a lip print.

This looked like a triumph for quantum theory, and the
Bohr-Sommerfeld model.

But...

Unfortunately Bohr and Sommerfeld were wrong. The
silver nucleus has no spin. What no one knew, until it was
proposed by Uhlenbeck and Goudsmit three years later,

was that electrons have spin. The silver atom has 23 pairs of electrons, and one lone electron on the outside. The spin of this lone electron was what was causing the beam to split. (All elements with odd atomic numbers have odd numbers of electrons – including hydrogen lithium, boron, nitrogen, fluorine, sodium – and silver.)

So the Stern-Gerlach experiment produced the correct result, although for the wrong reason. And yet the experiment was a success, for it was the first and most direct evidence in quantum mechanics of quantization – the fact that spin could have only two values.

Later, similar experiments showed that some atomic nuclei do have spin. In the 1930s Isidor Rabi showed that the spin could be pushed from one side to the other; this formed the basis of magnetic resonance imaging equipment used in hospitals. In the 1960s Norman F Ramsey modified the Rabi apparatus to make atomic clocks.

Even though Gerlach on his own performed the experiment, only Stern was awarded a Nobel Prize; apparently Gerlach was denied a share because of his later work in Nazi-led Germany. Nevertheless the Stern-Gerlach experiment is still hailed as one of the great experiments in quantum physics.

1923–1927

CAN A PARTICLE WAVE?

THE STUDY

RESEARCHERS:
Clinton Davisson
and Lester Germer

SUBJECT AREA:
Quantum mechanics

CONCLUSION:
Electrons are both
particles and waves.

PROOF OF WAVE-PARTICLE DUALITY

Surely things must be either particles or waves – or can they be both? In 1924 French physicist Louis de Broglie – actually Louis-Victor-Pierre-Raymond, 7th duc de Broglie – suggested in his PhD thesis that electrons behave as waves. He even outrageously suggested that all matter has wave properties. These suggestions were anathema to classical physicists, but quantum mechanics was sweeping forward, and perhaps his ideas had some validity. Most important, he derived an equation linking the energy of a particle to its wavelength.

After all, Einstein's 1905 paper on the photoelectron effect had shown that light comprised both waves and particles, bundled into photons, as we now call them. Could this apply to other things too? In Göttingen Walter Elsasser suggested that the wave nature of things might be investigated by scattering from crystalline solids.

Arthur Compton showed in 1923 by scattering x-rays from graphite that x-rays (and other forms of electromagnetic radiation) seem to have some mass; so they behave a bit like particles.

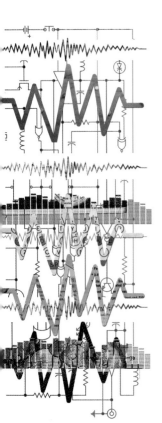

The experiment

In 1927, at Bell Labs in New Jersey, Clinton Davisson and Lester Germer wanted to find out about the surface structure of nickel metal, and planned to bombard it with a beam of electrons. They generated an electron beam by heating a wire filament, and then accelerated it with a moderate voltage, which they could vary in order to change the energy of the beam. At 50 V the electrons would have an energy of 50 electronvolts (eV).

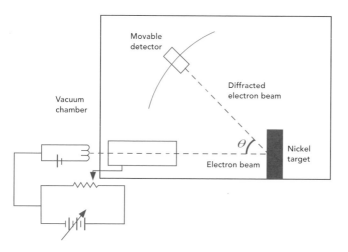

Movable
detector

Diffracted
electron beam

Vacuum
chamber

θ

Nickel
target

Electron beam

They then fired the beam at a right angle to the nickel
surface, and used a movable detector to measure the
angles of reflection. They expected that the rough surface
would scatter the electrons randomly in all directions,
and indeed this happened: 'Electrons... are scattered in
all directions at all speeds of bombardment', but after a lab
accident they found something quite unexpected.

The happy accident

The whole apparatus was contained in an evacuated
box, to prevent collisions with air molecules, but
unfortunately air leaked in, and the nickel was coated
with a layer of nickel oxide. So they heated the nickel to a
high temperature to get rid of the oxide. What Davisson
and Germer did not realize at the time was that the high
temperature altered the structure of the nickel, so that
the surface, which had been a mass of tiny crystals, was
now covered by a few large crystals, any one of which
was wider than the electron beam. As a result, when they
tried the experiment again, the beam was bouncing off a
single crystal of nickel.

Now they found that although some of the beam of
electrons was still scattered randomly, at particular
voltages many of the electrons came off at specific angles;
with an accelerating voltage of 54 V, for example, they

found a maximum in the deflected beam at an angle of 50 degrees. This was just like seeing a sudden flash of reflected sunlight from a window in a high building, or from the windshield of a distant automobile, when the angle is just right.

William Henry Bragg and his son William Lawrence Bragg had won the Nobel Prize in physics in 1915 'for their services in the analysis of crystal structure by means of x-rays'. They showed that x-rays bounced off crystals at particular angles, because crystals are built of layers of atoms, and when the angles are correct these layers can act as mirrors for the x-rays. Before long, x-ray diffraction was being used to work out crystal structures by measuring the angles and therefore calculating the distances between layers.

Particles and waves

Davisson and Germer report that when they reached particular voltages, sets of sharply defined beams of electrons flashed from the crystal in particular directions, with three or six beams in each set. Twenty of such sets of beams came off at the angles that they would have expected from beams of x-rays.

In other words, they discovered that electrons could act in just the same way as x-rays; so this means that electrons were behaving as waves.

Before this experiment, the electron had appeared simply to be a negatively charged particle, but now it had a wavelength. To some extent this was the opposite of the Compton effect, that waves of light have mass. What Davisson and Germer had shown was that particles can behave like waves, just as waves can behave like particles.

IS EVERYTHING UNCERTAIN?

HEISENBERG'S UNCERTAINTY PRINCIPLE

↓

1927
THE STUDY
RESEARCHER:
Werner Karl Heisenberg
SUBJECT AREA:
Quantum mechanics
CONCLUSION:
In the realm of the very small, we can't be absolutely certain about anything.

If we know how fast a particle is moving, we cannot simultaneously know where it is. German physicist Werner Heisenberg was one of the main pioneers of quantum mechanics. Born in Würtzburg, Germany, in 1901, he studied physics and mathematics at Munich and Göttingen. At the end of 1924 he went to work for Niels Bohr in Copenhagen, and it was there in 1927, while he was working out the mathematical basis of quantum mechanics, that he developed his uncertainty principle.

A thought experiment
He did not like the original model of quantum theory, with electrons in fixed orbits around the nucleus, because, he said, since he could not actually observe the electrons' orbits, people could not reasonably claim that electrons existed. All anyone could do was observe the light they emitted or absorbed, as they jumped from one orbital to another.

So he did a thought experiment (see page 95). Microscopes normally

use light waves to bring an image to the eye. Light from the sun or a lamp illuminates the specimen; some light bounces up the microscope tube, through the lenses and mirrors and into the observer's eye. Heisenberg wanted to observe an electron directly, but could not do so with light, because the wavelength of visible light is too long; it could not 'see' a tiny electron. That would be like trying to catch a speck of dust with a fishing net.

In order to get much higher resolution, he imagined that he had a microscope that used gamma rays instead of light waves. Gamma rays are like light waves, but have a very short wavelength, which should mean that the microscope would have super-high resolution; so he might be able to observe electrons directly, and find out where they were.

The problem

But a gamma ray has far more energy than a ray of light – so much energy that when it bounced off the electron it would undoubtedly give the electron a kick, and shove it away in an unknown direction. And if Heisenberg wanted to find out the position of the electron very precisely, he would need even higher-energy gamma rays, which would kick the electron even harder.

In other words, the more precisely he measured the position of an electron, the less information he would have

$$\Delta p \cdot \Delta q \approx h$$

about how fast it was moving, and in which direction. Conversely, the more precisely he could determine its trajectory, the less he would know about where it was.

Although he had discovered this idea by thinking about trying to measure the position of the electron, Heisenberg realized that the uncertainty was nothing to do with the method of measurement; it was an inherent property of the quantum world.

He explained his ideas in a letter to his friend Wolfgang Pauli, dated 23 February 1927. He developed a mathematical proof, and published a full paper the same year. The theory came to be called Heisenberg's uncertainty principle, and in due course formed part of the basis of the Copenhagen interpretation of quantum mechanics.

Never the same again

This uncertainty may not sound important, but in a small way it altered the whole of physics. Before that, in theory, if you knew the exact position and trajectory of a particle at one moment, you should be able to predict where it would be at any point in the future. This was the deterministic universe, as implied by Isaac Newton.

The Heisenberg uncertainty principle changed all that; for now he had shown that it was impossible to know both the position and the trajectory of a particle.

Luckily this applies only in the realm of quantum mechanics. In our 'real' world there are uncertainties, but they are too small to measure, or to matter. Newtonian physics got men to the moon, and they will continue to allow us to drive automobiles, and (with luck and skill) catch cricketballs.

1927–1929

RESEARCHERS:
Alexander
Alexandrovich
Friedman, Georges
Henri Joseph Édouard
Lemaître and Edwin
Powell Hubble

SUBJECT AREA:
Cosmology

CONCLUSION:
The universe began
with the Big Bang,
and is expanding ever
more rapidly.

WHY IS THE UNIVERSE EXPANDING?

THE COSMIC EGG

Alexander Friedman, a professor at Perm State University in Russia, suggested in a complex paper published in German in 1922 that the universe might be expanding.

The Belgian Catholic priest Henry Lemaître independently came to a similar conclusion, and produced a paper in 1927 on 'A homogeneous Universe of constant mass and growing radius accounting for the radial velocity of extragalactic nebulae'. In this paper he derived what came to be called Hubble's Law, and estimated what came to be called the Hubble constant.

Lemaître had worked as a graduate student with Arthur Eddington (see page 122) in Cambridge, England, and also in the US; so he was reasonably well known to English-speaking astronomers. Eddington in particular helped him to gain recognition, and to translate most of his paper into English.

Einstein was happy with Lemaître's mathematics, but did not believe the universe was expanding. Lemaître recalled Einstein saying '*Vos calculs sont corrects, mais votre physique est abominable.*' ('Your calculations are correct, but your physics is atrocious.')

In 1931 Lemaître published a paper in *Nature*:

I would rather be inclined to think that the present state of quantum theory suggests a beginning of the world very different from the present order of Nature. Thermodynamical principles from the point of view of quantum theory may be stated as follows:

(1) Energy of constant total amount is distributed in discrete quanta. (2) The number of distinct quanta is ever increasing. If we go back in the course of time we must find fewer and fewer quanta, until we find all the energy of the universe packed in a few or even in a unique quantum.

Lemaître suggested that the universe had expanded from a single point, and later spoke of 'the Cosmic Egg exploding at the moment of the creation.' Later, in a radio broadcast, the British astrophysicist Fred Hoyle, who did not believe that the universe was expanding, spoke dismissively of the 'Big Bang Theory', which is what it has been called ever since. Einstein eventually came round to Lemaître's point of view, and after hearing him speak at a seminar in California, said 'This is the most beautiful and satisfactory explanation of creation to which I have ever listened.'

Enter the American
The young Edwin Hubble, raised in Illinois and Kentucky, promised his father he would study law, and did so as one of the first Rhodes Scholars to go to Oxford, but his real love was astronomy, and he reverted to it after his father died.

After the end of World War I he spent a year in Cambridge, England, and then got a job at the Mount Wilson Observatory in Pasadena, California, where he remained for the rest of his life.

Hubble studied a special class of stars called Cepheid variables in various nebulae, including the Andromeda

Nebula. Cepheid variables are stars which get brighter and then dimmer in a regular cycle, lasting for some days. They are particularly interesting because there is a simple relationship between their brightness and the period of brightening and dimming. This means that from the period an astronomer can work out the absolute brightness of the star. For this reason they are known as 'standard candles'. Knowing how bright the standard candle really is, the astronomer can calculate, from its apparent brightness, how far away it is.

A nebula means a cloud of dust or gas. In the early 1920s all nebulae were thought to be clouds of dust or gas within our own galaxy, the Milky Way – and the Milky Way was thought to be the entire universe. Hubble showed that several galaxies were much, much further away than the most distant stars in the Milky Way – and therefore that these nebulae were actually distant galaxies. Suddenly he had shown that the universe was millions of time bigger than anyone had suspected.

The redshift

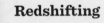

Redshifting

In 1929 Hubble looked at the redshift of 46 distant galaxies. It was known that when a galaxy (or a star) is moving away from us the light from it is 'redshifted' – that is, shifted towards the red (long wavelength) part of the spectrum.

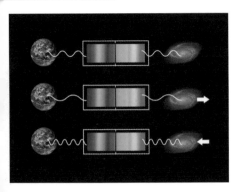

We now know that this is because space itself is being stretched, but the result is like the Doppler effect (see page 69). The more the redshift, the greater the speed of the galaxy away from us. Hubble found that the redshift of a galaxy was roughly proportional to its distance. In other words, the further away a galaxy was, the faster it was retreating.

DOES ANTIMATTER EXIST?

THE SEARCH FOR POSITIVE ELECTRONS AND NEGATIVE PROTONS

1932

THE STUDY

RESEARCHER:
Carl David Anderson

SUBJECT AREA:
Particle physics

CONCLUSION:
In addition to ordinary matter, antimatter also exists.

British theoretical physicist Paul Dirac, sometimes called the greatest theorist since Isaac Newton, developed his own strange form of mathematics, combining quantum mechanics with special relativity. He described the behaviour of an electron as it approached the speed of light – and then realized something odd. His 1928 equation was derived for a negatively charged electron, but would apply equally to a particle with a positive charge.

He went on to suggest that not only does the electron have an equivalent particle with the opposite charge, but so does every other particle. Just as protons and electrons combine to form atoms, so do anti-protons and anti-electrons combine to produce anti-atoms. In other words, he predicted the existence of antimatter, which had never been observed.

He even suggested that there might be complete solar systems of antimatter:

If we accept the view of complete symmetry between positive and negative electric charge so far as concerns the fundamental laws of Nature, we must regard it rather as an accident that the Earth (and presumably the whole solar system), contains a preponderance of negative electrons and positive protons.

It is quite possible that for some of the stars it is the other way about, these stars being built up mainly of positive electrons and negative protons. In fact, there may be half the stars of each kind. The two kinds of stars would both show exactly the same spectra, and there would be no way of distinguishing them by present astronomical methods.

So Dirac suggested that there may be antimatter stars and planets floating about in space.

Meanwhile, 15 years earlier, Austrian physicist Victor Hess had been interested in the amount of ionizing radiation in the atmosphere. This was thought to come from radioactive rocks in the Earth, but Hess calculated that if this were true, then the radiation should die out around 500 metres (1,640 feet) above the Earth's surface. So he decided to test this theory.

He made a series of ten balloon flights, at considerable risk, and found that the level of radiation decreased up to 1 kilometre (0.6 miles), but then increased again, and at 5 kilometres (3.5 miles) above Earth the radiation was twice as high as at sea level. He concluded that 'a radiation of very high penetrating power enters our atmosphere from above.'

He even made one ascent, in April 1912, at the time of a near-total eclipse of the sun, and found that the level of radiation did not decrease as the sun disappeared; so it could not have been coming from the sun. Hess had discovered what came to be called 'Hess rays', and later 'cosmic rays' – radiation from outer space; a stream of electromagnetic waves and particles that rains down on us day and night.

Anderson's cloud chamber

Carl Anderson studied physics and engineering at Caltech. In 1932 he began to study cosmic rays using a modified cloud chamber (see page 103). He called it a Wilson chamber, but later it became known as an Anderson chamber.

*On August 2, 1932, during the course of
photographing cosmic-ray tracks produced in a
vertical Wilson chamber ... the tracks shown ... were
obtained, which seemed to be interpretable only on
the basis of the existence in this case of a particle
carrying a positive charge but having a mass of the
same order of magnitude as that normally possessed
by a free negative electron.*

This is the key photograph of Anderson's cloud chamber, which has a 6 millimetre (0.24 inch) lead barrier across the middle. The cosmic ray is seen entering at the bottom, and curving to the left in the intense magnetic field; this proves it has a positive charge; with a negative charge it would have curved to the right. The particle then goes through the lead plate and comes out the other side with slightly diminished energy – which is why it then curves more sharply. The fact that it gets through the lead barrier and then through 5 centimetres (2 inches) of air proves that it is very small – a proton could not have travelled so far.

This was not an easy research task. Anderson had taken and examined 1,300 photographs, and found only 15 with similar positive cosmic ray tracks. 'It is concluded, therefore, that the magnitude of the charge of the positive electron which we shall henceforth contract to positron is very probably equal to that of a free negative electron.'

When an antiparticle collides with its counterpart – an electron with a positron, for example – they annihilate each other, giving off a gamma ray. Apparently there are no large regions of antimatter in space, for we have not observed the gamma-ray emissions that might be expected from collisions with normal matter. One of the great mysteries of cosmology is why the Big Bang produced more matter than antimatter.

Anderson's cloud chamber

14 cm (5.5 in) wide,
1 cm (0.4 in) deep

RESEARCHER:
Fritz Zwicky

SUBJECT AREA:
Cosmology

CONCLUSION:
There is much more
mass in the universe
than we can account
for in the visible stars.

HOW DOES GRAVITY
BIND GALAXIES?

DARK MATTER AND
THE MISSING UNIVERSE

Fritz Zwicky has been called one of the most brilliant
astrophysicists and one of the most unusual personalities
in the 20th century. How did this curmudgeonly genius
discover something invisible in space?

Zwicky was born in Bulgaria in 1898 to a Swiss father
and a Czech mother. He studied maths and physics
in Switzerland, and in 1925 emigrated to the United
States to work with Robert Millikan (see page 106)
at the California Institute of Technology. There he
became influential in the widening field of astronomy,
astrophysics and cosmology.

Supernovae and neutron stars

In the early 1930s Zwicky began to study 'new stars'
with the German astronomer Walter Baade. Zwicky had
conceived the notion that cosmic rays (see page 138) are
produced during the catastrophic explosion of a star, and
he called such an explosion a supernova. He and Baade
discovered 120 supernovae during the next 52 years. They
were not new – Tycho Brahe had observed one in 1572 –
but no one had explained them.

In 1933 Zwicky suggested that a typical massive star
ends its life in a colossal explosion, producing a burst of
bright light and cosmic rays. Left behind is a star so dense
that all the protons and electrons are crushed together
to form neutrons. The resulting neutron star would be
tiny, perhaps a few miles across, but unbelievably dense.

Neutrons had been discovered only the year before, and nobody really believed Zwicky, until Jocelyn Burnell discovered pulsars in 1967.

Zwicky had an extraordinary mind and capacity for lateral thinking. After he died, fellow astronomer Stephen Maurer wrote: 'When researchers talk about neutron stars, dark matter and gravitational lenses, they all start the same way: "Zwicky noticed this problem in the 1930s. Back then, nobody listened. . ."'

How much mass is there in a galaxy?
In 1932 the Dutch astronomer Jan Oort suggested that there must be more matter in the Milky Way than could be accounted for by observation, on the grounds of the movements of the stars, but his measurement was found to be wrong.

In 1933 Zwicky was the first scientist to apply the virial theorem to the Coma cluster of galaxies that lie 320 million light years away. The virial theorem provides an equation that describes the relationship between the orbital speeds of galaxies and the gravitational force acting on them.

Zwicky observed the movements of galaxies near the edge of the cluster, and from them estimated the total mass of the cluster. Then he estimated the mass from the number of galaxies and their masses, as implied by their brightness.

What he found was that the movements suggested about 400 times as much mass as he guessed from observation of the brightness. The mass of the material he could see was nothing like enough to cause the enormous speeds of the orbiting galaxies; something was missing. From this 'missing mass problem' Zwicky inferred that there must be a whole lot of invisible mass in the cluster. He called this *dunkle Materie*, or 'dark matter'.

Mysterious matter

In fact Zwicky was hugely inaccurate in his estimates, but the missing mass problem has not gone away, and astronomers have now found plenty of evidence to support his idea. In many, if not most, galaxies, the orbital speeds of the stars are much too great for the amount of visible massive material. It seems as though most galaxies have a roughly symmetrical sphere of dark matter with the visible stars in a disc at the centre.Observations of gravitational lensing (the effect was first suggested by Zwicky in 1937) have confirmed the presence of extra mass: a large concentration of matter, whether visible or dark, warps space-time (see page 97) so that objects far beyond may be observed as though through a lens, either magnified or distorted. This lensing is in some cases much more pronounced than can be explained by the visible matter alone.

In the late 1960s and early 1970s, Vera Rubin was able to measure the orbital velocities of stars in spiral galaxies, and found that most of the stars orbit at roughly the same speed, whereas those much further from the centre should orbit much more slowly. This suggests that the mass density of the galaxies remains much the same well beyond the central clump of stars and that most galaxies must contain about six times as much mass as can be explained by the visible stars.

Our own galaxy seems to have about ten times as much dark matter as visible matter. In 2005 astronomers from Cardiff University in Wales claimed to have discovered a galaxy with one-tenth of the mass of the Milky Way, and composed entirely of dark matter.

It is now thought that dark matter comprises around 27 per cent of the universe, with most of the rest accounted for by dark energy (see page 162).

IS SCHRÖDINGER'S CAT DEAD OR ALIVE?

THE PARADOX OF QUANTUM MECHANICS

1935

THE STUDY

RESEARCHER:
Erwin Schrödinger

SUBJECT AREA:
Quantum physics

CONCLUSION:
Two possibilities co-exist.

How can a cat be both alive and dead? This was the rhetorical question that Austrian physicist Erwin Schrödinger posed in 1935. For the previous 15 years various theoretical physicists and mathematicians had been working out the details of quantum mechanics. The principal architects of the theory were Niels Bohr and Werner Heisenberg, working in Copenhagen, and they produced what came to be called the Copenhagen interpretation of quantum mechanics (see page 131). Schrödinger thought there was a problem in the way that this applied to ordinary objects.

One theory Bohr and Heisenberg proposed was called 'quantum superposition'. When a particle (or a photon of light) could be in either of two states or positions, and there was no way of telling which, superposition said it was in both states at the same time, until it was actually

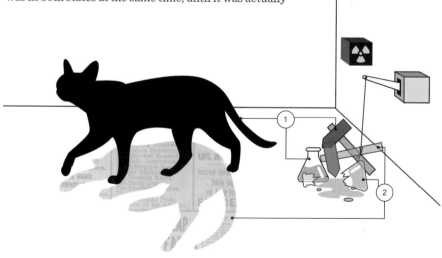

observed in one of them. Then it would instantly collapse into that state only. So only an observer could fix the particle in one of the possible states.

Schrödinger disliked the idea of superposition, and presented a paradox, in the form of a thought experiment.

Imagine a cat shut in a steel box with no means of escape. Beside it in the box is a mechanism containing a speck of radioactive material, a Geiger counter and a bottle of deadly cyanide poison. If one atom of the radioactive material decays, the Geiger counter will detect it, and activate a relay, causing a hammer to smash the bottle and release the poison, which will kill the cat.

Radioactive atoms are completely unpredictable. The one in the box might disintegrate after one second, or not for a year. Therefore, since no one can see into the box, no one can tell after say half an hour whether or not it has decayed. According to the superposition idea it should be both complete and disintegrated.

But that means that the cat should be both alive and dead – until an observer opens the box to find out. Schrödinger said this made no sense; superposition was ridiculous in the real world. He wrote that his paradox 'prevents us from accepting as valid a "blurred model" for representing reality. In itself, this would not embody anything unclear or contradictory.'

Some people argued that the cat was an observer; it would know whether the atom had exploded – while it was still alive.

Niels Bohr himself did not insist on the presence of the observer. For him the cat would be either alive or dead long before anyone opened the box. He reckoned that the Geiger counter would make the cat alive or dead. In effect the Geiger counter was an observer. Does this make any more sense? Not to Albert Einstein. He wrote to Schrödinger in 1950:

You are the only contemporary physicist... who sees that one cannot get around the assumption of reality, if only one is honest. Most of them simply do not see what sort of risky game they are playing with reality – reality as something independent of what is experimentally established. Their interpretation is, however, refuted most elegantly by your system of radioactive atom +... [mechanism]... + cat in a box... Nobody really doubts that the presence or absence of the cat is something independent of the act of observation.

Many worlds interpretation

Later models of quantum mechanics introduced different ideas. In 1957 Hugh Everett produced the 'many worlds interpretation', which suggested that when there are two possibilities, both remain true. Indeed, all possible histories and futures remain true. There is a large number of universes, and everything that could possibly have happened has happened, in one universe or another. In this theory, when Schrödinger's box is opened, both the observer and the alive/dead cat split in two. Then in one universe the observer has a live cat, while in the other the second observer has a dead cat – but the two observers can never meet or communicate with each other.

The arguments have carried on ever since, and Schrödinger's cat has become world famous – the most celebrated animal in quantum mechanics.

1939

THE STUDY

RESEARCHERS:

Leó Szilárd and
Enrico Fermi

SUBJECT AREA:

Nuclear physics

CONCLUSION:

Nuclear reactions can
generate energy.

HOW DID NUCLEAR PHYSICS LEAD TO THE A-BOMB?

THE FIRST NUCLEAR REACTOR

In 1933 Hungarian physicist Leó Szilárd was visiting England when he read in *The Times* newspaper about a speech given by Ernest Rutherford, by then a Grand Old Man of atomic physics. In the speech, Rutherford had dismissed the possibility of deriving energy from nuclear reactions: '. . .anyone who looked for a source of power in the transformation of atoms was talking moonshine.'

The frightening idea

This really annoyed Szilárd, and he was fuming about it as he went walking through London's Bloomsbury on the dull damp morning of 12 September. The story goes that he waited to cross Southampton Row, near the British Museum, until the traffic light turned red. Then, as he stepped off the pavement, the terrifying idea came to him: suppose someone could find a reaction that was started by the newly discovered neutron – a reaction in which an atom produced two neutrons. Then those two neutrons would start two more atoms reacting, to release four neutrons, and they

146

would produce eight... and there would be a chain reaction As Richard Rhodes put it in his book *The Making of the Atomic Bomb*: 'As he crossed the street time cracked open and he saw a way to the future, death into the world and all our woes, the shape of things to come.'

The brilliant Italian

Enrico Fermi was born in Rome, and had a sparkling career in physics, both theoretical and practical. In 1938 he won the Nobel Prize in physics for producing new elements by bombarding heavy atoms with neutrons. Unfortunately the 'new elements' turned out not to be new elements at all, but merely radioactive fragments generated in the reactions. Fermi was embarrassed, but remained full of confidence.

In 1939 war was threatening. Both Szilárd and Fermi had emigrated to the US to escape the Nazi and Fascist regimes. Realizing the possibility that German scientists might create an atomic bomb, they wrote a letter to warn President Roosevelt, and got Einstein to sign it too.

Critical mass

Meanwhile, other scientists had found that when a uranium atom disintegrates (see page 89) it generates two or three neutrons. Also, one slow neutron was found to be capable of causing a uranium atom to disintegrate. Here, then, was the possibility of a real nuclear chain reaction. Bring together a critical mass of uranium (about 15 kilograms or 33 pounds of pure metal – a lump a little bigger than a hockey ball) and the neutrons it produced would each cause further disintegrations – and the reaction would be unstoppable.

Fermi and Szilárd set about building the world's first nuclear reactor. They had come together to the University of Chicago, and planned to build the reactor in Red Gate Woods, safely outside the city, but a labor strike prevented

this; so they constructed 'Chicago Pile 1' (CP1) in a squash court underneath an abandoned sports centre, but still in the middle of a huge urban area.

The experiment was frighteningly dangerous. Szilárd, Fermi and others calculated that it would start and stop as planned, but if it had gone wrong the whole of Chicago would have been devastated. By then, however, America had been drawn into the world war, and perhaps the risk seemed justified.

Chicago Pile 1

The reactor was a pile of uranium pellets and graphite blocks. Fermi had discovered that the neutrons emitted when uranium atoms disintegrate are too fast to start a chain reaction. Paraffin wax or water slows them almost to a standstill, because the neutrons collide with all the hydrogen atoms in paraffin and water. Graphite was a more effective moderator, slowing the neutrons

just enough to make them highly effective at smashing another uranium atom.

They also needed a mechanism to slow down and stop the reaction – if it started; so they had an array of control rods made of cadmium and indium, ready to drop into slots in the pile. Cadmium and indium absorb neutrons; so they should be able to slow and stop the nuclear reaction.

The pile was assembled with the control rods in place, and then at 3:25 p.m. on 2 December 1942, the control rods were pulled back, and CP1 went critical. The first ever controlled nuclear reaction started. Fermi shut it down 28 minutes later.

Later the reactor was dismantled and moved to Red Gate Woods, where it became CP2, and later the first Argonne National Laboratory. Fermi went on to become a director at the Manhattan Project at Los Alamos, and measured the energy produced by the first atomic bomb test in the Alamogordo Desert in 1945.

CHAPTER 6: Across the universe: 1940-2009

In the early chapters of this book, scientists worked alone, building their own apparatus. As the work became more difficult and expensive, laboratories were constructed. Big Science has taken this several stages further.

Consider the advance of the tokamaks – doughnut-shaped machines for creating nuclear fusion. The first ones were made in Russia in Cold War secrecy, but gradually they grew, until the Joint European Torus in Britain produced a plasma with the highest temperature in the solar system. Soon, however, JET will be dwarfed by the colossal ITER experiment.

SuperWASP is a fine example of sheer ingenuity, coupled with massive computing power, but pride of place must go to the Large Hadron Collider, the biggest and most complicated piece of apparatus ever built.

In 1854 Louis Pasteur said: 'In the field of observation, chance favours the prepared mind.' A wonderful stroke of luck led to the discovery of the echo of the Big Bang in 1965, and two years later the same could be said for Jocelyn Bell's discovery of pulsars – luck and tenacity in equal parts, and this work catalyzed the exploration of black holes.

1956

THE STUDY

RESEARCHERS:
Igor Yevgenyevich
Tamm, Andrei
Dmitrievich Sakharov
and many others

SUBJECT AREA:
Nuclear physics

CONCLUSION:
Fusion may be
practical in the future.

A STAR IS BORN?

THE DEVELOPMENT OF
THE TOKAMAK

Nuclear fission has been used in reactors since the
1950s, but it remains an expensive energy source,
and because both starting materials and products are
radioactive, it carries various problems with it: dangers
of failure, meltdown, inundation by tsunami and
terrorist attack. Furthermore the long-term disposal
of radioactive waste is not simple. Nuclear fusion may
provide the answer.

Fusion and Fission
Fission means allowing massive atoms, such as those
of uranium or plutonium, to fall apart, releasing small
particles, atoms of lighter elements and a great deal
of energy. Fusion means slamming together two small
atoms, such as hydrogen, to make a larger one, such as
helium. There are several advantages. The system cannot
overheat and melt down, because the total mass of the
reacting gases at any one time will be less than a gram;
as a result, even though the materials are extremely hot,
the amount of heat is small – not enough to melt the steel
and ceramic walls. There is no problem about disposal
of the products, which are not fiercely radioactive, and
the reaction produces around a thousand times as much
energy as a typical fission reaction.

The sun's energy – and the energy of all the stars –
comes from fusion of hydrogen atoms to make helium; so
all we need to do is make a star on Earth. Both the physics
and the engineering of this, however, are extraordinarily
difficult. It has been said that a fusion reactor is only

30 years away – and always will be – but that has not
prevented people from trying.

The pioneers

The first scientists to get to grips with nuclear fusion were
Russians, and many details are not known, because they
started during the Cold War, when everything was secret.
We do know that Lev Artsimovich was a physicist, and had
been a member of the team working on the Soviet atomic
bomb. From 1951 until he died in 1973 he was head of the
Soviet fusion power project.

He led the team that produced the first
nuclear fusion reaction in any lab. When
he was asked when the first practical
thermonuclear reactor would start up,
he said 'When mankind needs it, maybe
a short time before that.' Artsimovich
was called 'the father of the tokamak'.

A tokamak is a reaction vessel
designed for nuclear fusion. The word
is an acronym; the original Russian
means 'toroidal chamber with magnetic
coils'. Imagine an inflated rubber ring
or automobile tyre. This shape, called a
torus, is the shape of the reaction vessel.

The first tokamaks were designed by Igor
Yevgenyevich Tamm and Andrei Dmitrievich
Sakharov, and built at the Kurchatov Institute
in Moscow in 1956, and the first successful fusion
reaction was achieved in Novosibirsk in 1968, with a
temperature of about 10 million °C (18 million °F). British
and American physicists confirmed this the following year.

There are now 30 tokamaks in operation in 16 different
countries. The largest so far is JET, the Joint European
Torus, at Culham, England, which has a tokamak easily
big enough for a man to walk around inside.

JET achieved its first plasma on 25 June 1983, and in 1997 produced 16 megawatts of fusion power, although for less than one second. But JET needed more power to keep going than it produced; so it was never going to be a commercial power station.

The magnetic bottle

In order to get hydrogen atoms to fuse into helium they have to be moving enormously fast, so that they slam into one another with great energy. To get them moving fast they must be heated to colossally high temperatures – like 100 million °C (180 million °F).

At these temperatures, the hydrogen has ceased to be a gas, and has become a plasma. This means that the molecules (H_2) have split into atoms (H·); then the electrons have come loose from the atoms, leaving protons (hydrogen ions H^+) whizzing around with the free electrons. These particles are both charged, which means that they can be contained in a 'magnetic bottle'.

If they were to crash into the wall of the reaction vessel they would lose much of their energy, and might also inflict serious damage on the wall; so they have to be restrained. This is done with immensely powerful magnetic fields, which travel round the inside of the torus and are also twisted, like the strands of a rope ring, so that they form spirals inside the torus. This complex magnetic field forms the magnetic bottle that keeps the hydrogen atoms away from the walls.

The energy generated by the fusion reaction is extracted mainly by the cooling water that flows through the double walls of the reaction chamber, but could also be taken out by the neutrons that are formed during the reaction, or by a process called direct energy conversion, which turns the speeding electrically charged particles directly into an electric current. This energy would be used to turn water into superheated steam, which would then drive turbines to make electricity, as in a conventional power station.

DID THE BIG BANG
HAVE AN ECHO?

DISCOVERING THE COSMIC
MICROWAVE BACKGROUND

1965
THE STUDY
RESEARCHERS:
Arno Allan Penzias
and Robert Woodrow
Wilson

SUBJECT AREA:
Cosmology

CONCLUSION:
We now have a map
of the young universe.

Arno Penzias was born in Munich, Germany, but left in
1939, and the family settled in New York. After gaining a
PhD in physics he went to work at Bell Labs in Holmdel,
New Jersey, with Texan physicist Robert Wilson.
Together they worked on a highly sensitive 15 metre (50
foot) microwave horn antenna/receiver, which had been
built in 1959 to make radio astronomy observations, and
to bounce signals off Echo balloon satellites. They hoped
to investigate radio signals from between galaxies.

Radio noise
When they switched the system on they heard radio
noise – that is a faint background hiss – which they
could not explain. They knew they would have to get rid
of this noise if they were to detect the faint signals they
were hoping for.

They eliminated all the effects of radio and TV
broadcasting; then to suppress interference from the heat
in the receiver, they cooled it with liquid helium to 4 K
(-269 °C or 500 °F). Still they heard the noise.

First they thought it must come from New York
City – from unsuppressed automobile spark plugs – so
they pointed the horn directly at

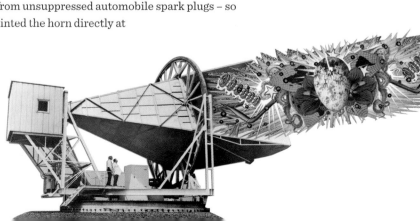

Manhattan, but the hiss got no louder. They realized it must be coming from the sky.

They suspected radiation from the Milky Way, but it was quieter than they would have expected, and more puzzling, it seemed to come from all over the sky, and was the same in every direction.

Pigeon droppings

Surely, they thought, it must be really local – possibly even coming from inside the receiver horn. They looked inside and found 'white dielectric material' – in other words, pigeon droppings. This might well explain the noise. So they cleared out the droppings, and then cleared out the pigeons roosts. And still the noise came hissing in.

Meanwhile at Princeton University, just 60 kilometres (37 miles) away, Robert Dicke and his colleagues Jim Peebles and David Wilkinson were just setting up to search for microwave radiation of this type; they predicted that it might have come from the Big Bang. Penzias heard from a friend who had seen a preprint of a paper by Peebles, and then he and Wilson realized how important their discovery was.

Penzias called Dicke, saw a copy of the paper and then invited Dicke and his colleagues over to Bell Labs to see their data. It seemed to fit the Princeton predictions well – 'We've been scooped,' said Dicke; so, in 1965, they published joint letters in the *Astrophysical Journal*.

The echo of the Big Bang

They were right; what Penzias and Wilson had heard was the cosmic microwave background radiation (CMB), which is in effect the echo of the Big Bang.

The Big Bang released an unimaginable burst of energy into the universe, some of which eventually condensed into matter. When the universe was only about 380,000 years old, it became transparent, and this energy must have seemed like a billion simultaneous continuous lightning flashes, with a colour temperature around 3,000 K. Colour temperature is used by photographers to describe whiteness; 500 °C or 932 °F (773 K) is red hot; 1,500 °C or 2,732 °F is yellow; 3,000 K (2,727 °C or 4,940 °F) is white; sunlight is about 5,000 K (4,727 °C or 8,540 °F).

The universe aged, and after 13.7 billion years got to where we are today. This ancient light is still flooding the universe, but space has expanded so fast, the light has been redshifted (or cooled) all the way to the microwave region of the spectrum. So we are now seeing, as microwaves, the oldest light in the universe – the remains of the light from the Big Bang. The microwaves have a wavelength of 7.3 centimeters (2.9 inches), which corresponds to black-body radiation at a temperature of 3 K (three degrees above absolute zero).

This discovery provided powerful support for the Big Bang theory, which was still in contention with the steady-state theory. Big Bang theory predicted CMB – and Penzias and Wilson found it.

A map of the young universe

Although Penzias and Wilson thought it was isotropic – that is, exactly the same in all directions – it turns out to be very slightly lumpy; the temperature varies from 3 K by less than one thousandth of a degree, but it does vary. This picture is actually a map of the universe as it was at the age of only 380,000 years, 13.77 billion years ago.

The yellow and especially the red patches show where the light was more intense. These are the places where matter began to clump together to form stars and eventually galaxies. This is the best map we have of the young universe.

1967

THE STUDY

RESEARCHER:
Susan Jocelyn Bell

SUBJECT AREA:
Astronomy

CONCLUSION:
Black holes are real.

DO LITTLE GREEN
MEN EXIST?

PULSARS AND BLACK HOLES

How were black holes discovered? On 26 May 1783, British clergyman and polymath John Michell (see page 57) wrote a long letter to Henry Cavendish at the Royal Society. In it he described a sphere 500 times bigger than the sun:

> *a body falling from an infinite height toward it, would have acquired at its surface a greater velocity than that of light, & consequently, supposing light to be attracted by the same force... all light emitted from such a body would be made to return towards it, by its own proper gravity.*

In other words, Michell had conceived the idea of a black hole, a body so massive that even light could not escape its gravitational attraction. Just 13 years later the same idea appeared in the book *Exposition du système du Monde*, by French mathematician Pierre-Simon Laplace.

The idea of the black hole was resurrected after Einstein's 1915 paper on general relativity had ignited a new fire in cosmology. German physicist Karl Schwarzschild found a solution to Einstein field equations, specifying the gravitational field of a point mass and a spherical mass. Something strange occurred at what became known as the Schwarzschild radius; this has now become known as the 'event horizon' – a spherical shell through which matter can enter but nothing can leave.

So black holes existed in mathematics, but could they exist in reality?

The PhD student

In 1967 astronomer Jocelyn Bell, born in Northern Ireland, was a graduate student at Cambridge University, England. Her mission was to look for quasars, mysterious new astronomical objects, and her first task was to build a radio telescope by stringing miles of wires on wooden posts – 'I became very good at wielding a sledgehammer', she said.

After getting rid of local interference from badly suppressed cars and thermostats, she was able to get on with her work and look for quasars, but then she noticed an unexpected signal – a weak bit of 'scruff' on her chart paper. After several times riding six miles on her scooter to the observatory in the middle of the night, she managed to get a clear magnified recording of it – a series of radio pulses exactly 1.337 seconds apart.

Bell and her supervisor, Antony Hewish, thought such a regular signal must be man-made, but then she discovered that it was coming from the sky, and from a particular point in the sky.

For a while she thought it must be an alien civilization, and she called it LGM-1, for Little Green Men. Then, just before Christmas, she found another signal, LGM-2, this time with peaks 1.25 seconds apart. Surely there could not be two alien civilizations trying to make contact?

Eventually they found out that the signals were coming from neutron stars, which had been predicted in 1934, but never seen. Neutron stars seem to be formed after the collapse of massive stars. They are made entirely of neutrons, without any electrons to keep them apart, and so are extraordinarily dense; a neutron star with a diameter 12 kilometres or 7 miles would have twice the mass of our sun.

These ones were spinning rapidly and sending out radio beams that swept across the universe like the beams of light from a lighthouse. They became known as pulsars. Bell discovered the first four, and now more than 2,000 have been found.

Antony Hewish (but not Jocelyn Bell) won a Nobel Prize in 1974.

Could black holes be real?

Once neutron stars were known to be real, interest in black holes was revived, because astrophysicists knew that gravitational collapse was possible. Black holes cannot be observed directly, because they emit virtually no light, although Stephen Hawking worked out that they emit a very weak infrared signal. However, their presence can be inferred by their effects on their neighbours; some stars orbit black holes, for example.

There may even be a supermassive black hole at the centre of every galaxy, including our own. Observation of the trajectories of 90 stars near the centre of the Milky Way suggests the presence of a black hole with a mass 2.6 million times that of our sun.

There seem to be several sizes of black holes: micro holes, with mass about the same as our moon's, some as massive as the sun and supermassive black holes, millions of times bigger. Black holes seem to be formed when supermassive stars collapse. Small stars turn into neutron stars, but bigger ones have so much mass that the immense gravity squashes even the neutrons into a point, called a singularity.

IS THE UNIVERSE ACCELERATING?

OUR LONELY FUTURE

1998

THE STUDY

RESEARCHERS:
Saul Perlmutter,
Adam Riess,
Brian P Schmidt, et al

SUBJECT AREA:
Cosmology

CONCLUSION:
The universe is
expanding faster
and faster.

Gravity is the only force acting on all the galaxies today, and although gravity is weak over large distance, it is relentless. Gravity should be pulling everything together again, so that the expansion of the universe should eventually slow down and reverse; having started with a Big Bang, the universe should end with a Big Crunch. But when?

Saul Perlmutter set up the Supernova Cosmology Project (SCP) in 1998 with 20 collaborators in the US, Europe and Chile, to find out how quickly the expansion was slowing down, and to predict when the big crunch would happen.

Brian P Schmidt and Adam Guy Riess were part of the Hi-Z Supernova Research Team (HZT) at the Australian National University's Mount Stromlo Observatory. 'High-Z' means high redshift; that is, supernovas enormously distant from Earth. The HZT project, like the SCP, wanted to find out how much the expansion of the universe was slowing.

The plan of both teams was to measure the distance and redshift of galaxies enormously distant from Earth. Distance and redshift are related by Hubble's Law; the further away an object is, the more its light is shifted to the red end of the spectrum. Because the light from far-distant objects takes billions of years to reach us, the comparison of their distances and redshifts would tell the researchers how fast the universe was expanding in the distant past.

They needed to find extremely distant 'standard candles' – objects of known brightness that would allow the astronomers to work out their distances, from their apparent brightness as seen from Earth. They chose supernovas of a particular type, called IA supernovas, which seem to appear when a white dwarf star gains too much matter from a companion star and explodes.

Astonishing results

Later in 1998 both teams published papers, with similar astonishing results. They compared the distances of these supernovas with their redshifts, expecting them to follow Hubble's Law, or to be moving faster than Hubble's Law would predict.

To their surprise both teams found that the distant supernovas showed significantly smaller redshifts than predicted by their distances. This means that billions of years ago, when the light of the explosions left them, the galaxies were moving more slowly than predicted by Hubble's Law.

In other words, the expansion has accelerated.

Dark energy

'Dark energy' or 'vacuum energy', appears to act as a small negative pressure – a vacuum pulling the universe apart.

Cosmologists tell us that dark energy permeates space, and causes the galaxies to accelerate outwards, expanding the universe faster and faster. They now say that the universe comprises roughly 5 per cent ordinary matter, 27 per cent dark matter and 68 per cent dark energy.

WHY ARE WE HERE?

LIFE, THE MULTIVERSE AND EVERYTHING

1999
THE STUDY
RESEARCHERS:
Martin Rees, Stephen Hawking and others

SUBJECT AREA:
Cosmology

CONCLUSION:
We still have more questions than answers.

Why are we here? The question has puzzled philosophers and scientists for thousands of years. In his book *Just Six Numbers*, published in 1999, British Astronomer Royal Martin Rees defines six numbers that 'constitute a "recipe" for a universe... if any one of them were to be 'untuned', there would be no stars and no life. Is this tuning just a brute fact, a coincidence? Or is it the providence of a benign Creator?'

Rees says that neither of these possibilities is correct, but that there may be an infinite number of other universes where the numbers are different. They may have different laws of physics; they may have different chemical elements or different properties of atoms; they may have no small molecules that could evolve into life. We could have evolved only in a universe where the numbers are 'right'.

In their book *The Grand Design*, Stephen Hawking and Leonard Mlodinow use the analogy of bubbles in a pan of water coming to the boil. Many small bubbles appear on the bottom of the pan, and then collapse again, representing universes that don't last long enough to develop stars and galaxies, let alone intelligent life. Some bubbles do survive, however, and grow big enough to rise to the surface and release steam – and they represent universes that do develop.

The anthropic principle

The idea that our universe is just right for us is part of what is known as 'the anthropic principle'. The Greek word *anthropos* means 'human'. The strong anthropic principle says that this universe was somehow compelled to take the form that allows human beings to evolve.

One version of the weak anthropic principle suggests that of many possible universes, we inhabit the one that has all the right characteristics – the correct values of Rees's six numbers.

The phrase 'anthropic principle' was coined by Brandon Carter in 1973, but the idea goes back more than 100 years. Alfred Russel Wallace was the man who narrowly beat Charles Darwin in writing down the idea of evolution by means of natural selection. In 1904 he wrote:

> *Such a vast and complex universe as that we know*
> *exists around us, may have been absolutely required*
> *...in order to produce a world that should be*
> *precisely adapted in every detail for the orderly*
> *development of life culminating in man.*

Rees uses the analogy of a large coat shop. If you go in and find that they have an enormous and varied stock, then you will probably find one that fits. Likewise if there were not just one Big Bang but many Big Bangs, then you might reasonably expect one of them to produce a universe with the characteristics that suit us. So perhaps there are dozens, hundreds or even millions of other universes out there.

The quantum world
A photon of light seems to be able to take two paths in order to go through two different slits (see page 64). Richard Feynman said this is because in the quantum world the photon does not have a unique history, but instead takes every possible path. It may follow that when the universe began, it set off in every possible way, creating a range of universes, most of which are quite

unlike ours. This is closely related to Hugh Everett's 'Many worlds' interpretation of quantum mechanics. If Schrödinger's cat is alive and dead in different worlds (see page 143), they may be different universes – but that means that the observer who opens the box creates a new universe by doing so.

Furthermore our own universe is almost unimaginably vast. Our own galaxy contains 200 billion stars, most of which appear to have planets. Beyond our galaxy are at least 100 billion other galaxies, each with its own complement of stars, and (probably) planets. This seems an awful lot of stuff to be created just for us, and would the observer really create an equivalent amount just by opening the box?

If there are other universes, why can't we see them?

A colony of ants, running about on a two-dimensional sheet of paper, may not know that a few centimetres above them is another sheet of paper, with another colony of ants. This top sheet is another universe, but separated from the first by a third dimension, to which the ants have no access. In the same way there may be another universe in another dimension, to which we have no access. It may be only inches away, but we are not aware of it. One complex theory of physics, called 'M-theory', suggests there are eleven dimensions – plenty to accommodate other universes.

On the other hand, if we cannot in any way interact with any of these other universes, why even imagine that they exist? Occam's razor suggests that we should always look for the simplest explanation for any phenomenon; therefore, perhaps we should abandon the idea of fictitious universes.

2007

THE STUDY

RESEARCHERS:
Don Pollacco et al

SUBJECT AREA:
Astronomy

CONCLUSION:
There are many
habitable exoplanets
in our galaxy.

ARE WE ALONE IN THE UNIVERSE?

WASP AND SUPERWASP

On 6 October 1995, Swiss scientists Michel Mayor and Didier Queloz, working at the Haute-Provence observatory in southeast France, announced that they had discovered a planet in another solar system. Its official name is 51 Pegasi b.

This was the first discovery of an exoplanet orbiting an ordinary star. It's an enormous planet, bigger than Jupiter and is unusually close to its star, completing an orbit in just over four days. They found it because its gravitational attraction rocks the parent star to and fro, which means that the colour of the star shows a periodic Doppler shift (see page 69).

Could there be life out there?
Once exoplanets were known to exist, astronomers began hunting for them in earnest. If we are not alone in the universe, then life is probably to be found on a planet like Earth – a small, rocky planet with a temperature in the 'Goldilocks zone', (not too cold and not too hot) between about 0 °C (32 °F) and 60 °C (140 °F), so that it could have liquid water on the surface.

A major problem for planet hunters is that planets don't shine. Stars are easy to see, but planets are small and dark, and are generally invisible against their bright stars.

Don Pollacco and his colleagues at Queen's University, Belfast, Northern Ireland, found a simple way to search. They reckoned that there might be many exoplanets out there, and that occasionally one of them, while orbiting its star, would pass in front of the star, blocking some of its light. Therefore they could simply look at the stars, and watch out for slight periodic dimming, which might mean a planet was crossing in front.

Digital cameras

This ingenious group bought four high-tech digital cameras with 200 millimetre (7.9 inch) f/1.8 Canon lenses. With help from Cambridge University, Instituto de Astrofisica de Canarias, and the Isaac Newton Group of Telescopes, they mounted them in a small fibreglass shed on a mountaintop on La Palma, in the Canary Islands, off the Western Sahara. They called their project the Wide Angle Search for Planets, or WASP. Then Queen's and the Open University provided more funds; they bought four more cameras, and renamed the project SuperWASP; it was up and running in 2002.

There was a slight problem, however. Canon had stopped making the 200 millimetre lenses to match the first four; so Pollacco bought them on eBay. The eight cameras are all mounted on one robot arm, and are arranged so that they diverge slightly, and therefore between them cover a wide angle of sky.

Pictures of the stars

All eight cameras take two pictures, with different exposure times; then the robot arm swings over towards a new patch of sky and they take two more, and so on until they have covered the entire sky. Then they return to the starting position. During the night the cameras take some 600 photographs. Each image contains up to 100,000 stars.

SuperWASP identifies every star by comparison with an astronomical catalogue. Then it measures the brightness of each star. After several months, the researchers search for dips in brightness that might indicate a planet passing in front of the star.

The most obvious dimming is caused by very large planets, and is easiest to spot if it happens frequently – that is, if the planet is close to and rapidly orbiting its star, so that the dimming occurs every few days. These planets are called 'hot Jupiters' (like 51 Pegasi b), and are fairly common, but not promising for life; they are much too hot to have liquid water on the surface, and the gravitational effects would be impossibly powerful for life.

Exoplanets galore
SuperWASP announced its first exoplanet WASP-1 in 2007; it's a hot Jupiter with an orbital period of only 2.5 days. WASP-12b is so close to its star that it is heated to around 1,500 °C (2,800 °F) and is pulled by huge gravitational forces into shape of a football. By 2015 SuperWASP had found more than 100 exoplanets.

Inspired, perhaps, by the success of SuperWASP, NASA launched in 2009 a spacecraft called *Kepler*, which looks continuously at 145,000 stars, watches for dimming and has found more than a thousand exoplanets, plus 3,000 possibles.

Astronomers now reckon that most stars probably have their own planetary systems, and that in the Milky Way alone there may be as many as eleven billion rocky, Earth-like planets within the Goldilocks zone. Surely 'in some warm little pond' as Darwin put it, on one of those 11,000,000,000 planets, something might be alive.

CAN THE HIGGS BOSON BE FOUND?

THE LARGE HADRON COLLIDER

2009
THE STUDY

RESEARCHERS:
Peter Higgs and
12,000 scientists
from 100 countries

SUBJECT AREA:
Particle physics

CONCLUSION:
It is likely that the
Higgs boson has
been found.

Particle physicists spend their time trying to understand the elementary particles that make up atoms. Most of us are content with protons, electrons and neutrons, but they have unearthed an entire zoo of smaller particles, from neutrinos to quarks – particle physicists have created a new language too – and for several decades they have fitted them into what is called the 'standard model'.

In 1964 Peter Higgs, at the University of Edinburgh, Scotland, predicted that within the standard model there should be a particle which gives other particles mass. This particle should be a boson – but no one could find it, and it became 'the most sought-after particle in modern physics'.

Colliders

The faster particles are travelling the more damage they can do – and the more secrets may be revealed. So physicists have invented various ways of speeding up the particles to enormous velocities. First they used electrostatic accelerators, then linear accelerators (linacs) in which a succession of electric fields each give the particles a kick forward.

As a bunch of particles approach a plate they are attracted by the opposite electrical charge on it, but as they zoom through a hole in the plate the charge is switched, so that it now repels them and makes them go faster towards the next plate – and this process continues down the line.

Next came the cyclotron, which is like a linac but bent into a circle. The particles are pulled into the circular path by electromagnets; they go round and round, getting faster all the time, until they reach an energy of around 15 million electronvolts (eV). The synchrotron is a more advanced cyclotron, in which the guiding magnetic field is synchronized with the particle beam.

The Large Hadron Collider

A hadron is, technically, a particle made of quarks held together by the strong force. Protons – the nuclei of hydrogen atoms (H⁺) – are hadrons, and the Large Hadron Collider (LHC) was built primarily to accelerate hadrons, especially protons, using a linac and synchrotrons.

About 328 ft (100 m) below the border between France and Switzerland is a nearly circular tunnel, about 4 metres (12 foot) wide and 27 kilometres (17 miles) long, housing a pair of pipes about 10 centimetres (4 inches) in diameter. Inside each pipe a stream of protons hurtles along, one stream clockwise round the circuit, the other anticlockwise. These pipes comprise an enormous synchrotron.

Before getting into the pipes, the protons have been accelerated by a linac and by three successive synchrotrons, and within the pipes they are accelerated even more, over a period of 20 minutes, until they reach at 99.999999 per cent of the speed of light – just 3 metres per second (9.8 feet per second) slower than light speed, which gives them an energy of around 4 million million electronvolts (compare the 50 eV used by Davisson and Germer). Each proton goes round the 16.7 mile (27 km) circuit 11,000 times every second.

The proton beams are steered around the circuit and focused by a total of 1,600 superconducting magnets, each weighing nearly 30 tons, and all cooled to 1.9 K (-271 °C or -456 °F) by 96 tons of liquid helium.

There are four crossing points in the ring, where the two pipes become one, so that the protons careering one way can smash into those coming the other way. This is where reactions happen, and the crossing points are surrounded by detectors to observe the debris.

When the collider is operating at full power, there are millions of collisions every second, and each one produces streams of particles, automatically observed in the detectors, almost like high-tech cloud chambers (see page 104). This produces a staggering amount of data, which is fed for analysis via a dedicated computing grid to 170 computers in 36 countries.

The first collisions took place on 23 November 2009, and within a few months the LHC was running at full power.

Physicists hoped to find out whether the Higgs boson really exists and also to tackle some of the big unknowns of particle physics – to look for the mysterious 'dark matter' which seems to make up 25 per cent of the universe, and for new particles predicted by theories of 'supersymmetry' – an extension of the standard model that aims to fill some gaps.

So far, the researchers have discovered several new composite hadron particles. They have observed a quark-gluon plasma, which is thought to be what the universe comprised up to a few milliseconds after the Big Bang. They have seen one rare particle decay, which seems to be evidence against supersymmetry. And above all they have seen evidence for the elusive Higgs boson.

The LHC has spoken.

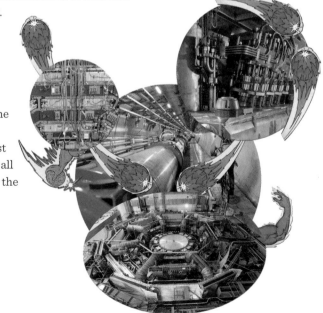

Index

acoustics (studies), 69–71
Alexander the Great, 17
al-Hakim, Caliph, 20
Alhazen, 9, 20–22
Ampère, André-Marie, 66
Anderson, Carl, 115, 137–39
Arago, François, 66
Archimedes, 6, 8, 13–16, 19
Aristotle, 8, 21, 32–33, 35
Artsimovich, Lev Andreevich, 152–54
Aston, Francis William, 88
astronomy (studies), 158–60, 163–65
astrophysics (studies), 122–24
atomic physics (studies), 86–88, 98–100, 119–21, 125–27
Avogadro, Amedeo, 103

Baade, Walter, 140
Bacon, Francis (philosopher), 26
Banks, Joseph, 61–62
Becquerel, Antoine Henri, 83–85, 91
Bell, Susan Jocelyn, 151, 158–60
Bertha, Anna, 84
Black, Joseph, 49–51
Blackett, Patrick, 121
Bohr, Niels, 111–12, 121, 125–26, 134, 143, 144
Bonaparte, Napoleon, 61
Boyle, Robert, 6, 11, 27, 37–39, 78, 80, 105
Bradley, James, 75
Bragg, William Henry, 133
Bragg, William Lawrence, 133
Brahe, Tycho, 140
Burnell, Jocelyn, 141

Carter, Brandon, 164
Cassini, Giovanni Domenico, 43–45
Cavendish, Henry, 57–59, 158
Clausius, Rudolf, 141
Compton, Arthur, 131, 133

Conduitt, John, 48
Copernicus, Nicolaus, 6
cosmology (studies), 80–82, 134–36, 140–42, 155–57, 161–62, 163–65
Crookes, William, 78, 83
Curie, Marie Skłodowska, 79, 89–91
Curie, Pierre, 89–91

Daguerre, Louis-Jacques, 75
Darwin, Charles, 76, 164, 168
Davidson, Charles, 122–24
da Vinci, Leonardo, 22
Davisson, Clinton, 115, 128–30, 170
Davy, Humphry, 62, 67–68
de Broglie, Louis, 115, 131
Descartes, René, 22, 48
Dicke, Robert, 156
Dirac, Paul, 115, 137–38
Doppler, Christian Andreas, 69–71, 130, 166
Dyson, Frank Watson, 122–24

Eddington, Arthur S, 122–24, 128
Edison, Thomas, 92, 103
Einstein, Albert, 95–97, 104, 113, 114, 116–18, 122–24, 125, 128, 134, 135, 144–45, 147, 158
studies, 95–97, 116–18
electricity (studies), 60–62, 92–94, 101–2
electromagnetic spectrum (studies), 83–85
electromagnetism (studies), 66–68
Elsasser, Walter, 131
Empedocles, 6, 8, 10–12
Eratosthenes, 8, 17–19
Euclid, 21
Everett, Hugh, 145, 165

Faraday, Michael, 53, 66–68, 74, 83, 125
Fermi, Enrico, 146–49
Feynman, Richard, 65, 164–65
Fizeau, Hippolyte, 71, 75–77
Fletcher, Harvey, 106–9
Foucault, Léon, 75–77
Franck, James, 110–13
Franklin, Benjamin, 60
Friedman, Alexander Alexadrovich, 134–36

Galileo, Galilei, 6, 22, 26–27, 31–33, 34, 37, 43, 96, 116
Galvani, Luigi, 60–61
Geiger, Johannes Wilhelm, 98–100
Geissler, Heinrich, 83
general relativity (studies), 116–18, 122–24
geometry (studies), 17–19
Gerlach, Walther, 125–27
Germer, Lester, 115, 128–30, 170
Gilbert, William, 30
Goldstein, Eugen, 83
Goudsmit, Samuel, 126
gravity (studies), 31–33, 54–56
Guericke, Otto von, 27

Hafele, Joseph, 117
Halley, Edmond, 45, 46–47, 48
Hawking, Stephen, 160, 163–65
Heisenberg, Werner Karl, 115, 131–33, 143
Hertz, Gustav Ludwig, 110–13
Hess, Victor, 138
Hewish, Antony, 159–60
Higgs, Paul, 169–71
Hooke, Robert, 37–39, 46, 48
Hoyle, Fred, 129
Hubble, Edwin, 71, 134–36, 161
Hutton, Charles, 56
Huygens, Christiaan, 32, 63
hydrostatics (studies), 13–16

JET (tokamak), 150, 153
Joule, James Prescott, 52,
    72–74, 105

Keating, Richard, 117
Kelvin, William Thomson,
    Baron, 105–6, 114

Laplace, Pierre-Simon, 158
Large Hadron Collider (LHC),
    106, 151, 169–71
Lavoisier, Antoine, 72
Lemaître, Henri, 134–36
Lenard, Philip, 83, 85

Marsden, Ernest, 98–100
Maskelyne, Nevile, 54–56, 58
Mason, Charles, 54–55
Maurer, Stephen, 141
Mayor, Michel, 166
Michell, John, 57, 58, 59, 158
Michelson, Albert A, 80–82, 96
Millikan, Robert Andrews,
    106–9, 140
Minkowski, Hermann, 97
Mlodinow, Leonard, 163
Moore, Gordon, 7
Morgan, John Pierpont, 94
Morley, Edward W, 80–82, 96

Napoleon I, Emperor of France,
    61
Newton, Isaac, 6, 22, 27, 33, 39,
    45, 54, 56, 63, 64, 76, 118, 123,
    124, 133, 137
studies, 40–42, 46–48
Norman, Robert, 26–27, 28–30
nuclear physics (studies),
    146–49, 152–54

Ockham, William of, 165
Onnes, Heike Kamerlingh,
    101–2
Oort, Jan, 141
optics (studies), 20–25, 40–42,
    43–45, 63–65, 75–77
Ørsted, Hans Christian, 66–68

particle physics (studies),
    101–4, 107–9, 137–39, 169–71
Pascal, Blaise, 27, 34–36, 37
Pasteur, Louis, 151
Pauli, Wolfgang, 135
Peebles, Jim, 156
Penzias, Arno Allan, 155–57
Périer, Florin, 35
Perlmutter, Saul, 161–62
Planck, Max, 112
pneumatics (studies), 10–12,
    37–39
Pollacco, Don, 166–68
Pound, Robert, 117
Power, Henry, 38
Ptolemy, Claudius, 21

quantum mechanics (studies),
    110–13, 128–30, 131–33,
    134–36
quantum physics (studies),
    143–45
Queloz, Didier, 166

Rabi, Isidor, 127
radioactivity (studies), 83–85,
    89–91
Ramsey, Norman F., 127
Rayleigh, Robert John Strutt,
    Baron, 82
Rebka, Glen, 117
Rees, Martin, 163–65
Rhodes, Richard, 146–47
Riess, Adam, 161–62
Rømer, Ole, 43–45, 75
Röntgen, Wilhelm Conrad,
    83–85, 114
Rubin, Vera, 142
Rumford, Benjamin Thompson,
    Count, 72
Rutherford, Ernest, 79, 91,
    98–100, 109, 119–21, 125, 146

Sakharov, Andrei Dmitrievich,
    153
Schmidt, Brian P., 161–62
Schrödinger, Erwin, 143–45
Schuster, Arthur, 86

Schwarzschild, Karl, 158
Sommerfeld, Arnold, 126
Sprengel, Hermann, 78
Stern, Otto, 125–27
Stukeley, William, 47–48
SuperWASP (Wide Angle
    Search for Planets), 150,
    166–68
Szilárd, Leó, 146–49

Tamm, Igor Yevgenyvich, 153
Tesla, Nikola, 92–94
Thales (mathematician), 10
Theodoric of Freiberg, 9, 23–25
thermodynamics (studies),
    49–51, 72–74
Thompson, Benjamin, 72
Thomson, Charles, 103–5
Thomson, Joseph John, 86–88,
    98, 101
Thomson, William, 74
Tokamaks, 150, 152–54
Torricelli, Evangelista, 27,
    34–35, 37, 39, 78, 80
Towneley, Richard, 27, 38–39
Townsend, John S, 86–88

Uhlenbeck, George, 126

Volta, Alessandro, 60–62

Wallace, Alfred Russel, 164
WASP see SuperWASP
Watt, James, 51, 58
Westinghouse, George, 92
Wilkinson, David, 156
Wilson, Harold A, 86–88
Wilson, Rees, 103–5
Wilson, Robert Woodrow,
    155–57
Wollaston, William Hyde, 67–68
Wren, Christopher, 46

Young, Thomas, 63–65, 77

Zhang Heng, 8
Zwicky, Fritz, 140–42

Glossary

alpha particle – the nucleus of a helium atom, comprising two protons and two neutrons.

blueshift – decrease in wavelength, or increase in frequency.

cathode rays – electrons streaming from the cathode in a vacuum.

dark matter – invisible matter that appears to constitute 84.5 per cent of the total mass of the universe.

event horizon – the boundary of a black hole; anything can go in, but nothing can come out, not even light (though black holes do emit small amounts of radiation, called Hawking Radiation).

exoplanet – planet outside our solar system – not in orbit around our Sun.

inertial frame of reference – a place that is stationary, or moving in one direction at a constant speed, but not accelerating.

m-theory – an idea in particle physics, developed from 'string theory', which attempts to explain all the particles and all the energy in the universe.

photoelectric effect – the emission of electrons by metals when light shines on them.

photon – a unit of light energy; a packet of light waves.

plasma – the three main states of matter are solid, liquid, and gas. Plasma is a fourth state in which all the particles are ionized (fire, for example, is a plasma).

polyphase – A system of distributing alternating-current electrical power using three or more electrical conductors.

positron – an antimatter particle: like an electron, but with a positive charge.

redshift – increase in wavelength, or decrease in frequency.

scintillation – flashes of light when particles hit a phosphorescent screen.

SI units – International system (Système Internationale) of units of measurement.

spectrometer – instrument for measuring the spectrum of an atom.

spin – in quantum mechanics, the angular momentum of a particle.

superposition – in the Copenhagen interpretation of quantum mechanics, the idea that a particle can be in two or more places at once.

supersymmetry – an extension of the Standard Model of particle physics, which predicts a partner for every particle.

thermocouple – a device for measuring temperature, made of two different metals joined at one spot.

uranium – a heavy metallic element which is radioactive.

Acknowledgements

I would like to thank Silvia Langford for inviting me to write about so many old friends, Slav Todorov and especially Sir Michael Berry for helping me survive Special Relativity, and my former colleagues Paul Bader, Marty Jopson and John Francas for introducing me to so many ancient scientists.

Sources

Chapter 1 Kingsley, Peter. *Ancient Philosophy, Mystery and Magic: Empedocles and Pythagorean Tradition* (Oxford, UK: Oxford University Press, 1995).

'On Floating Bodies' in *The Works of Archimedes*, ed. Heath, T. L., Cambridge, 1897 (New York: Dover Publications, 2002).

Chambers, James T. 'Eratosthenes of Cyrene' in Magill, Frank N. ed., *Dictionary of World Biography: The Ancient World* (Pasadena, CA: Salem Press, 1998).

Sabra, A. I., ed., *The Optics of Ibn al-Haytham* (Kuwait: National Council for Culture, Arts and Letters, 1983, 2002).

Harré, Rom. *Great Scientific Experiments: 20 Experiments that Changed our View of the World* (Oxford UK: Phaidon, 1981).

Chapter 2 Norman, Robert. *The Newe Attractive* (London: Ballard, 1581).

Galilei, Galileo. *Discorsi e Dimostrazioni Matematiche Intorno a Due Nuove Scienze* (Leiden: Louis Elsevier, 1638).

Pascal, Blaise. *Experiences nouvelles touchant le vide (New experiments on the vacuum)* (1647).

Boyle, Robert. *New Experiments Physico-Mechanical: Touching the Spring of the Air and their Effects* (1660).

Newton, Isaac. *Philosophical Transactions of the Royal Society of London* 6 (1671/2): 3075–3087.

(Rømer, Ole. Never officially published.)

Newton, Isaac. *Philosophiae Naturalis Principia Mathematica (The mathematical principles of natural philosophy)* (London, 1687).

Derham William. 'Experimenta & Observationes de Soni Motu, Aliisque ad id Attinentibus (Experiments and Observations on the speed of sound, and related matters).' *Philosophical Transactions of the Royal Society of London* 26 (1708): 2–35.

Black, Joseph. Lecture, April 23, 1762, University of Glasgow.

Chapter 3 Maskelyne, Nevil. 'An Account of Observations Made on the Mountain Schehallien for Finding Its Attraction. By the Rev. Nevil Maskelyne, BDFRS and Astronomer Royal.' *Philosophical Transactions of the Royal Society of London* (1775): 500–542.

Cavendish, Henry. 'Experiments to Determine the Density of the Earth. By Henry Cavendish, Esq. FRS and AS.' *Philosophical Transactions of the Royal Society of London* (1798): 469–526.

Volta, Alessandro. Letter to Sir Joseph Banks, March 20, 1800. 'On the Electricity Excited by the Mere Contact of Conducting Substances of Different Kinds.' *Philosophical Transactions of the Royal Society of London* 90 (1800): 403–431.

Young, Thomas. 'The Bakerian lecture: On the theory of light

and colours.' *Philosophical Transactions of the Royal Society of London* (1802): 12–48.

Cayley, George. 'Sir George Cayley's governable parachutes.' *Mechanics Magazine*, September 25, 1852.

Faraday, Michael. 'On some new electro-magnetical motions, and on the theory of magnetism.' *Quarterly Journal of Science* 12 (1821).

Doppler, Christian Andreas. 'On the coloured light of the double stars and certain other stars of the heavens.' *Abh. Kgl. Böhm. Ges. d. Wiss.* (Prague) (1842): 465–482.

Joule, James Prescott. 'On the Mechanical Equivalent of Heat.' *Abstracts of the Papers Communicated to the Royal Society of London* (1843): 839–839.

Fizeau, Hippolyte, and Léon Foucault .'Méthode générale pour mesurer la vitesse de la lumière dans l'air et les milieux transparents. Vitesses relatives de la lumière dans l'air et dans l'eau' (General method for measuring the speed of light in air and transparent media. Relative speed of light in air and in water.) *Compt. Rendus* 30 (1850): 551.

Bessemer, Henry. *Sir Henry Bessemer – FRS, An Autobiography* (London: The Institute of Metals, 1905).

Chapter 4 Michelson, Albert A., and Morley, Edward W. 'On the Relative Motion of the Earth and the Luminiferous Ether.' *American Journal of Science* 34 (1887): 333–345.

Röntgen, W. C. 'Über eine neue Art von Strahlen' (On a New Kind of Rays). *Sitzungsberichte der Würzburger Physik-medic. Gesellschaft* (1895).

Thomson, Joseph John. 'XL. Cathode rays.' *The London, Edinburgh, and Dublin Philosophical Magazine and Journal of Science* 44, no. 269 (1897): 293–316.

Curie, P. and Curie, M. S. 'Sur Une Nouvelle Substance Fortement Radio-Active, Contenue Dans La Pitchblende' (On a new radioactive substance contained in pitchblende). *Comptes Rendus* 127 (1898): 175–8.

Tesla, Nikola. *Colorado Springs Notes 1899–1900* (Beograd: Nolit, 1978).

Einstein, Albert. 'Zur Elektrodynamik bewegter Körper.' *Annalen der Physik* 17 (1905): 891.

Geiger, Hans, and Ernest Marsden. 'LXI. The laws of deflexion of ☐ particles through large angles.' *The London, Edinburgh, and Dublin Philosophical Magazine and Journal of Science* 25, no. 148 (1913): 604–623.

Onnes, H. Kamerlingh. 'The disappearance of the resistivity of mercury.' *Comm. Phys. Lab. Univ. Leiden*; No. 120b, 1911. Proc. K Ned. Akad. Wet. 13, (21911) 1274.

Wilson, Charles Thomson Rees. 'On a method of making visible the paths of ionising particles through a gas.' *Proceedings of the Royal Society of London. Series A, Containing Papers of a Mathematical and Physical Character* 85, no. 578 (1911): 285–288.

Franck, J. and Hertz, G. 'Über Zusammenstöße zwischen Elektronen und Molekülen des Quecksilberdampfes und die Ionisierungsspannung desselben' (On the collisions between electrons and molecules of mercury vapor and the ionization potential of the same).

Verhandlungen der Deutschen Physikalischen Gesellschaft 16 (1914): 457–467.

Chapter 5 Einstein, Albert 'Die Feldgleichungen der Gravitation' (The Field Equations of Gravitation). *Königlich Preussische Akademie der Wissenschaften*. 1915: 844–847.

Rutherford, Ernest. 'LIV. Collision of alpha particles with light atoms. IV. An anomalous effect in nitrogen.' *The London, Edinburgh, and Dublin Philosophical Magazine and Journal of Science* 37, no. 222 (1919): 581–587.

Dyson, Frank W., Arthur S. Eddington, and Charles Davidson. 'A determination of the deflection of light by the sun's gravitational field, from observations made at the total eclipse of May 29, 1919.' *Philosophical Transactions of the Royal Society of London: A Mathematical, Physical and Engineering Sciences* 220, no. 571–581 (1920): 291–333.

Gerlach, W., and O. Stern. 'Der experimentelle Nachweis der Richtungsquantelung im Magnetfeld.' *Zeitschrift für Physik* 9 (1922): 349.

Friedman, Alexander. '*Über die Krümmung des Raumes.*' *Zeitschrift für Physik* 10 (1922): 377–386.

Lemaître, Georges. 'Un Univers homogène de masse constante et de rayon croissant rendant compte de la vitesse radiale des nébuleuses extra-galactiques.' *Annales de la Société Scientifique de Bruxelles* 47 (1927): 49.

Hubble, Edwin. 'A relation between distance and radial velocity among extra-galactic nebulae.' *Proceedings of the National Academy of Sciences* 15, no. 3 (1929): 168–173.

Davisson, Clinton, and Lester H. Germer. 'Diffraction of electrons by a crystal of nickel.' *Physical review* 30, no. 6 (1927): 705.

Heisenberg, Werner. 'Über den anschaulichen Inhalt der quantentheoretischen Kinematik und Mechanik.' *Zeitschrift für Physik* 43, no. 3–4 (1927): 172–198.

Anderson, Carl D. 'The positive electron.' *Physical Review* 43, no. 6 (1933): 491.

Schrödinger, Erwin. 'Die gegenwärtige Situation in der Quantenmechanik (The present situation in quantum mechanics).' *Naturwissenschaften* 23 (49) (1935): 807–812.

Chapter 6 Fermi, E. 'The Development of the first chain reaction pile.' *Proceedings of the American Philosophical Society* 90 (1946): 20–24.

Bondarenko, B. D. 'Role played by O. A. Lavrent'ev in the formulation of the problem and the initiation of research into controlled nuclear fusion in the USSR.' *Phys. Usp.* 44 (2001): 844.

Penzias, Arno A., and Robert Woodrow Wilson. 'A Measurement of Excess Antenna Temperature at 4080 Mc/s.' *The Astrophysical Journal* 142 (1965): 419–421.

Hewish, Antony, S. Jocelyn Bell, J. D. H. Pilkington, P. F. Scott, and R. A. Collins. 'Observation of a rapidly pulsating radio source.' *Nature* 217, no. 5130 (1968): 709–713.

Cameron, A. Collier, F. Bouchy, G. Hébrard, P. Maxted, Don Pollacco, F. Pont, I. Skillen et al. 'WASP-1b and WASP-2b: two new transiting exoplanets detected with SuperWASP and SOPHIE.' *Monthly Notices of the Royal Astronomical Society* 375, no. 3 (2007): 951–957.

Rees, Martin. *Just Six Numbers* (London, Weidenfeld & Nicolson, 1999).

Gianotti, F. ATLAS talk at 'Latest update in the search for the Higgs boson.' CERN, July 4, 2012. Incandela, J. CMS talk at 'Latest update in the search for the Higgs boson.' CERN, July 4, 2012. Aad, Georges, T. Abajyan, B. Abbott, J. Abdallah, S. Abdel Khalek, A. A. Abdelalim, O. Abdinov et al. 'Combined search for the Standard Model Higgs boson in p p collisions at s= 7 TeV with the ATLAS detector.' *Physical Review D* 86, no. 3 (2012): 032003.